Springer Geology

The book series Springer Geology comprises a broad portfolio of scientific books, aiming at researchers, students, and everyone interested in geology. The series includes peer-reviewed monographs, edited volumes, textbooks, and conference proceedings. It covers the entire research area of geology including, but not limited to, economic geology, mineral resources, historical geology, quantitative geology, structural geology, geomorphology, paleontology, and sedimentology.

More information about this series at http://www.springer.com/series/10172

Francisco J. Prevosti · Analía M. Forasiepi

Evolution of South American Mammalian Predators During the Cenozoic: Paleobiogeographic and Paleoenvironmental Contingencies

 Springer

Francisco J. Prevosti
Centro Regional de Investigaciones
  Científicas y Transferencia Tecnológica
  de La Rioja (CRILAR), Provincia de La
  Rioja, UNLaR, SEGEMAR, UNCa,
  CONICET
Anillaco, La Rioja
Argentina

and

Departamento de Ciencias Exactas,
  Físicas y Naturales
Universidad Nacional de La Rioja (UNLaR)
La Rioja, La Rioja
Argentina

Analía M. Forasiepi
IANIGLA, CCT-CONICET-Mendoza
Mendoza
Argentina

ISSN 2197-9545   ISSN 2197-9553 (electronic)
Springer Geology
ISBN 978-3-319-79139-5   ISBN 978-3-319-03701-1 (eBook)
https://doi.org/10.1007/978-3-319-03701-1

This Springer imprint is published by Springer Nature
The registered company is Springer International Publishing AG
The registered company address is: Gewerbestrasse 11, 6330 Cham, Switzerland

Preface

Understanding of the patterns and processes of the evolution of the South American fauna during the Cenozoic presents an exciting challenge. South America hosted an extraordinary biota during the last 65 million years, including entire clades that radiated and then vanished, leaving no counterparts in recent times. Any analysis of this history is reliant on the fossil record, with its inherent limitations and biases, and as the record improves over time, any interpretation will be subject to review. New scientific techniques and statistical analysis throw also fresh light onto existing ideas.

In this work, we examined the relationship between two interesting groups of South American mammalian predators. The now extinct clade of the native Sparassodonta (distant relatives to current marsupials) found themselves confronted with the incoming Carnivora in relation to the establishment of the Panama Bridge linking both Americas. We look at their possible paleobiological interactions and examine some possible hypothesis of competitive displacement.

Proposed in the second half of the eighteenth century, the theory of dispersalism— the definition of a center of origin, usually a small area, and consequent dispersal to other areas of colonization—historically prevailed as an explanation of the geographic distribution of organisms. Initially, the center of origin was considered to be a tropical island (the "Garden of Eden"), but later migrated in idea to a location in the Northern Hemisphere. The role of the Southern Hemisphere was seen as a home of preserved relics of "ancient biota" and the receptor of the "more competitive" biota coming from the north (the "Sherwin Williams" effect). Although a great deal of work has refuted several concepts and hypotheses, our understandings still have some affection for them.

Dispersalism provided the commonly held view that the mammalian predator immigrants to South America from the Northern Hemisphere (the placental Carnivora) were "better adapted" than the native Sparassodonta. This was a convenient explanation for the replacement by the Carnivora of the Sparassodonta and subsequent extinction of the latter group. With better data, new scientific methods, and fresh ideas, this simplistic view has been challenged.

In this contribution, we present a review of the information available from the fossil record and examine the possible interactions and outcomes of the meeting of members of the two groups. We are privileged to be able to call upon more than 20 years of scientific research made by the authors on South American mammalian predators. Older works were highly influential and were integrated in the discussion, as well as the researches made by other workers from the last decades on connected subjects, such as systematic, biogeography, paleoecology, geology, and biostratigraphy. Much of the work is of a technical nature, but hopefully we have succeeded in our wish to make the contribution accessible to the non-specialist reader.

Anillaco, La Rioja, Argentina   Francisco J. Prevosti
Mendoza, Argentina   Analía M. Forasiepi

Acknowledgements

We are very grateful for the assistance of a number of colleagues who have helped in reviewing parts of the work and taking the time for instructive and informative conversations. Our debt extends to Lars Werdelin, Ross MacPhee, and Eduardo Tonni for the helpful comments given after the read of the manuscript of the book or parts of it and Luis Borrero, Fabiana Martin, Sergio Vizcaíno, Guillermo Cassini, Juan C. Fernícola, Susana Bargo, Francisco Goin, Judith Babot, Catalina Suarez, Natalia Zimicz, Amelia Chemisquy, Mariano Ramírez, Sebastian Echarri, Mauro Schiaffini, and Ulyses Pardiñas for their comments, data and support during these years. To Claudia Tambussi for good advices and to encourage to submit this contribution. Jorge Rabassa and Springer team for work editing the manuscript. For compiling published information for the construction of the datasets, many thanks to Sebastian Echarri and Sergio Tarquini who helped with references of Chaps. 1, 2, and 4. Mauro Schiaffini helped with calculation of geographic area covered by the localities of each South American Age.

We are very grateful to the curators and staff who assisted during collection visits: Alejandro Kramarz, Stella Álvarez, Marcelo Reguero, David Flores, Itatí Olivares, Mario Trindade, Diego Verzi, Alejandro Dondas, Martín de los Reyes, Fernando Scaglia, Itatí Olivares, M. Trindade, Diego Verzi, Alejandro Dondas, Fernando Scaglia, Matias Taglioretti, Richard Tedford, John Flynn, Bruce MacFadden, Richard Hulbert, Bill Simpson, Pablo Ortíz, Gabriel Acuña Suarez, Héctor Arzani, Ross D.E. MacPhee, Masanaru Takai, Shintaro Ogino, Ursula Göhlich, Barbara Herzig, Doris Nagel, Doris Moerike, Reinier van Zelst, R. Vonk, Christine Argot, Bruce Patterson, Ascanio Rincón, Judy Galkin, Min-Tho Schulenberg, William Stanley, Walter Joyce, Linda Gordon, Géraldine Veron, Matthew Carrano, Sumru Arincali, Tom Amorosi, Luciano Prates, Mariano Bonomo, Alfredo Prieto, Guillermo Delia, Amador Rodríguez, Daniel Ibáñez, D. Dias Henriques, Alejandro Salles, Fabiana Martin and José Luis Carrion, Bill Simpson, Pablo Ortíz, Gabriel Acuña Suarez, Héctor Arzani. We thank Sergio F. Vizcaíno and Susana Bargo for granting access to sparassodont specimens from the exciting recent excavations from the Santa Cruz Formation (fieldwork financially supported by PICT2013-389 to S.F. Vizcaíno).

FJP wants to express his deepest gratitude and appreciation to Amelia, Ursula and Chiara for constant support. Also to his parents (Alicia and Roberto) and sisters (Carla and Marisa) for their support throughout life.

Simon Kay has reviewed early versions of the text to mitigate errors in the English. AMF expresses enormous gratitude for his support. Ross MacPhee patiently reviewed the last version of the manuscript still finding many mistakes. FJP and AMF are deeply grateful for his assistance and time.

Jorge Blanco contributed with beautiful reconstructions of the extinct mammals and landscapes. Daniel Dueñas helped us with the drawings of the maps in Chaps. 1 and 2. Some pictures were generously provided by Lars Werdelin (Fig. 4.6b), Mauro Schiaffini (Fig. 4.10a), Z. Jack Tseng and Lorraine Meeker (Fig. 4.3a).

Visits to various institutions and particular projects were enabled by study collection grants and other travel grants supported by CONICET, DAAD, JSPS, American Museum of Natural History, Florida Museum of Natural History, Field Museum of Natural History, Fulbright Commission, and SNSF-IZK0Z3_160806. Our research has been supported by CONICET and ANPCyT (Argentina) grants PIP2011-164, PICT2011-309 and PICT2015-966.

Contents

Chapter 1
Introduction

Abstract During most of the Cenozoic, South America was an "island continent," sporadically connected with other landmasses. This feature resulted in the development of a peculiar biota in which endemic South American taxa were mixed with immigrants from other continents. The mammalian taxonomic diversity of South America was mainly composed of two groups: Metatheria and Eutheria. In this scenario, the carnivorous adaptive zone in South America was represented principally by metatherians (Sparassodonta). However starting in the late Miocene, this guild began to be occupied by placental carnivores (Eutheria, Carnivora), which, by the start of the Pleistocene, had become the dominant terrestrial predators. The changes in the ecosystems during the late Neogene were related to sparassodont extinction and subsequent placental immigration, in the context of the Great American Biotic Interchange. This book summarizes paleontological information about the origin, systematics, phylogeny, paleoecology, and evolution of the Sparassodonta and Carnivora, the two mammalian carnivorous groups in South America, including hypotheses about the interaction between these clades.

Keywords Mammalia · Carnivores · Cenozoic

1.1 Mammalian Carnivores

Many kinds of vertebrates with a carnivorous diet have existed, and mammals in particular have produced impressive radiations in this adaptive zone (Van Valen 1971; Marshall 1978; Martin 1989; Werdelin 1996; Van Valkenburgh 1999). Mammals demonstrate the complete range of carnivory, from omnivory (eat vertebrates, invertebrates, and food from vegetal origin in a similar proportion) to hypercarnivory, and in exploiting this niche have demonstrated convergent evolution, even within families or between larger taxonomic groups (Van Valkenburgh 1991, 1999; Fig. 1.1).

The placental Order Carnivora (more than 250 living species) includes living and fossil relatives of dogs, cats, bears, hyaenas, weasels, otters, skunks, raccoons,

© Springer International Publishing AG 2018   1
F.J. Prevosti and A.M. Forasiepi, *Evolution of South American Mammalian Predators During the Cenozoic: Paleobiogeographic and Paleoenvironmental Contingencies*, Springer Geology, https://doi.org/10.1007/978-3-319-03701-1_1

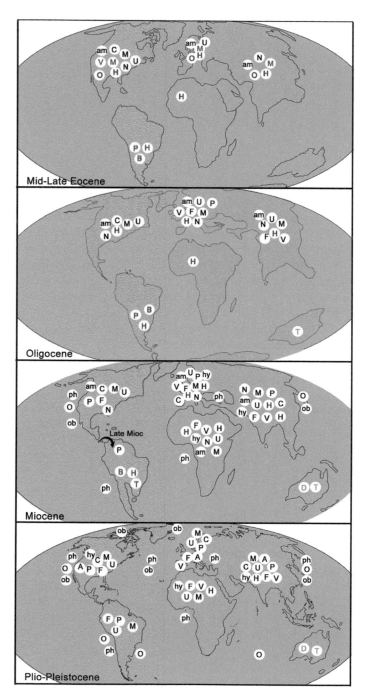

◄**Fig. 1.1** Geographic distribution of mammalian carnivores through the Cenozoic. "CREODONTA" (blue): H, Hyaenodontidae; O, Oxyaenidae. CARNIVORA (black): A, Ailuridae; am, Amphicyonidae; C, Canidae; F, Felidae; H, Herpestidae; hy, Hyaenidae; M, Mustelidae, Mephitidae; N, Nimravidae; O, Otariidae; ob, Odobenidae; P, Procyonidae; ph, Phocidae; U, Ursidae; V, Viverridae. Stem Carnivora in violet: M, Miacidae; V, Viverravidae. SPARASSODONTA (red): B, Borhyaenidae; H, Hathliacynidae; P, Proborhyaenidae; T, Thylacosmilidae. DASYUROMORPHA (green): D, Dasyuridae; T, Thylacinidae (adapted from Flynn and Wesley-Hunt 2005)

civets, mongoose, and seals (Figs. 1.2–1.3). Carnivorans include a wide range of body shapes, body masses, and locomotor types and are found in most habitats around the world, from the tropics to the poles and in the seas (Ewer 1973; Nowak 2005; Wilson and Reeder 2005; Wilson and Mittermeier 2009; Chap. 4).

The Order Carnivora can be traced back to the early Paleocene ($\sim$66–61.6 Ma) in North America and the late Paleocene ($\sim$61.6–56 Ma) in Asia (Hunt 1996; Flynn and Wesley-Hunt 2005). Evidence from Africa shows their presence from the late Oligocene (27–5.33 Ma), and from South America in the late Miocene ($\sim$7.3 Ma) (Patterson and Pascual 1972; Simpson 1980; Webb 1985; Hunt 1996; Flynn and Wesley-Hunt 2005; Rasmussen and Gutierrez 2009; Werdelin 2010; Woodburne et al. 2006; Woodburne 2010; Prevosti and Soibelzon 2012; Prevosti et al. 2013).

This clade has shown extensive dietary adaptation, with some members secondarily evolving to omnivorous (e.g., raccoons) or even herbivorous (e.g., giant pandas) diets. These specializations are manifest in the morphology of the dentition and skull (Crusanfont-Pairó and Truyols-Santoja 1956; Van Valkenburgh 1989; Biknevicius and Van Valkenburgh 1996; Friscia et al. 2007; Jones 2003; Evans and Fortelius 2008; Meloro and O'Higgins 2011; Figueirido et al. 2011, 2013; Prevosti et al. 2012a, b).

Carnivorans were not alone in eating their neighbors in the Cenozoic terrestrial ecosystems: "Creodonta" (Fig. 1.1) currently divided in Hyaenodonta and Oxyaenodonta but now considered paraphyletic, were the dominant predatory mammals of the northern continents during the Paleogene (Flynn and Wesley-Hunt 2005; Fig. 1.1). Hyaenodonta was present in Africa since the Paleocene and became extinct in the latest middle Miocene or earliest late Miocene (Lewis and Morlo 2010) and persisted in Laurasia until the middle Miocene (Flynn and Wesley-Hunt 2005). Oxyaenodonta were restricted to northern continents and became extinct in the middle Eocene (Solé and Smith 2013; Flynn and Wesley-Hunt 2005). "Creodonts" had a much lower diversity and anatomical disparity than Carnivora today (Werdelin 1996; Van Valkenburgh 1999; Flynn and Wesley-Hunt 2005).

Some of the greatest specializations for a carnivorous diet are seen in marine carnivorans (Fig. 1.3) with homodont dentitions—all teeth with the same shape—which is a very unusual feature among placental groups. Shrews, bats, and the

CREODONTA

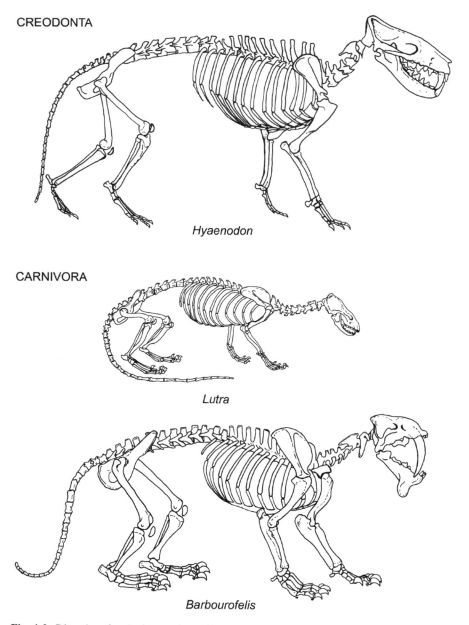

Hyaenodon

CARNIVORA

Lutra

Barbourofelis

Fig. 1.2 Diversity of eutherian carnivore (Creodonta and Carnivora) body plans (modified form Ewer 1973)

extinct groups Mesonychidae and Entelodontidae also have or had carnivorous adaptations but were limited in the degree of dietary specialization or diversification (Freeman 1984; Werdelin 1996; Van Valkenburgh 1999; Peigné et al. 2009).

CARNIVORA

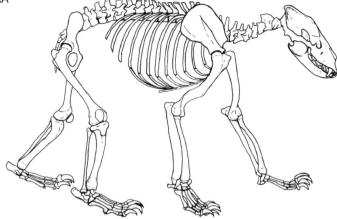

Ursus

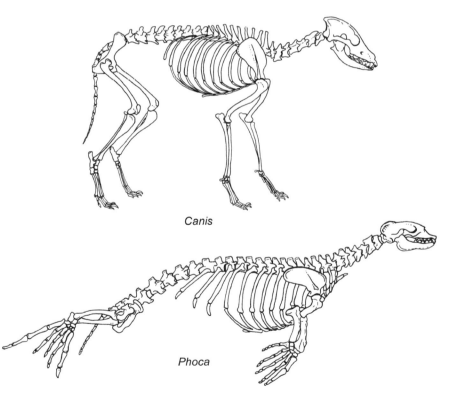

Canis

Phoca

Fig. 1.3 Diversity of carnivoran (Carnivora) body plans (modified form Ewer 1973)

In the southern hemisphere, several clades of marsupials and their fossil relatives
have occupied the carnivorous adaptive zone in the "island continents" of the
Cenozoic (Australia and South America; Fig. 1.1), producing some striking cases of
evolutionary convergence. These include, for example, the sabretooth sparassodont
—*Thylacosmilus atrox*—that was analogous to the sabretooth placental felids, or
the marsupial wolves—*Thylacinus cynocephalus*—that have been compared with
true (placental) wolves (Werdelin 1987; Jones and Stoddart 1998; Jones 2003;
Wroe and Milne 2007; Wroe et al. 2007; Figueirido and Janis 2011). In Australia
and Tasmania, the marsupial Order Dasyuromorphia radiated to produce more than
60 species of mostly small carnivorous mammals (BM < 1 kg), with some larger
exceptions like the marsupial wolf (*Thylacinus cynocephalus*, BM 15–35 kg), the
scavenger Tasmanian devil (*Sarcophilus harrisii*, BM 7–9 kg), and several intermediate species (e.g., *Dasyurus maculatus*, BM 4–7 kg) (Strahan 1995; Jones
2003; Wilson and Reeder 2005). The Australian carnivorous specialists also
included the extraordinary Thylacoleonidae, animals related to kangaroos
(Diprotodontia). The marsupial lion (*Thylacoleo carnifex*) was a large (BM ~
100 kg) predator that could have hunted megammamals like *Diprotodon* (Wroe
et al. 2003) and became extinct in the Late Pleistocene. Another marsupial lineage
that apparently occupied the carnivore role in Australia was the extinct
Propleopinae kangaroos of the Miocene–Pleistocene, with a body mass between 2.5
and 47 kg (Wroe et al. 1998; Wroe 2002).

In South America, the Sparassodonta—an extinct stem marsupial clade—were
diverse, but less so than extant carnivorans, with mesocarnivores (species that eat
mainly small vertebrates, but also some significant amount of invertebrates or
plants) to highly specialized hypercarnivores (Fig. 1.4), including bone crackers
and sabretooth forms and having body masses from less than 1 kg to over 100 kg
(Marshall 1978; Argot 2004; Forasiepi 2009; Ercoli and Prevosti 2012; Prevosti
et al. 2012a, b; Ercoli et al. 2014; Babot and Forasiepi 2016; Chap. 3). With a range
presumably extending from the Paleocene, to their last recorded appearance in the
Pliocene (~3.5 Ma), their extinction was originally explained by competitive
displacement by North American placentals (Simpson 1950, 1969, 1971, 1980;
Patterson and Pascual 1972; Savage 1977; Werdelin 1987; Wang et al. 2008).
Recent studies now discount the hypothesis with ecological replacement as the
likely process (Forasiepi et al. 2007; Prevosti et al. 2013; Chap. 6). Some didelphimorphians gave rise to carnivorous species during the late Miocene and
Pliocene, but compared to Sparassodonta and Carnivora, their diversity was low,
with species of small size (BM < 1–7 kg) that were not as dentally specialized
toward carnivory (Goin 1989; Goin and Pardiñas 1996; Prevosti et al. 2013; Zimicz
2014; Goin et al. 2016).

This work attempts to present a description and to discuss the evolution of
mammalian carnivores in South America. Chapter 2 is a discussion of climatic,
environmental, tectonic, and faunal changes in the Cenozoic. Chapters 3 and 4

SPARASSODONTA

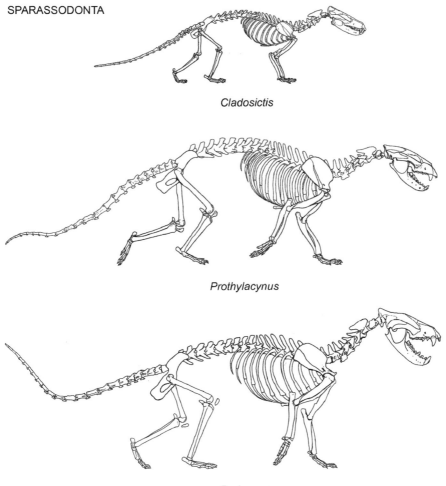

Cladosictis

Prothylacynus

Borhyaena

Fig. 1.4 Diversity of metatherian carnivore (Sparassodonta) body plans

address the systematics, phylogeny, and paleoecology of Sparassodonta and Carnivora. Chapter 5 explores the South American fossil record of these carnivores, together with limitations and biases (see Table 1.1 for the South American Age scheme used in this work and the time limits of each Age). Chapter 6 analyzes changes to the South American carnivore guild from the extinct sparassodonts to the extant carnivorans. We examine how these clades could have interacted during the late Miocene–Pliocene and also the potential impact of changing environmental and tectonic events, and the effect of other fauna in the evolution of South American carnivores.

Table 1.1 South American Age used in this book with their time limit and references

Ages	Time limits	References
Recent	1492 AD– present	Cione et al. (2015)
Platan	7 ka–1492 AD	Cione et al. (2015)
Lujanian	126–7 ka	Cione et al. (2015)
Bonaerian	400–126 ka	Cione et al. (2015)
Ensenadan	1.78–0.4 Ma	Cione et al. (2015)
Sanandresian	2.7–1.78 Ma	Bidegain and Rico (2012); Prevosti et al. (2013)
Vorohuean	2.9–2.7 Ma	Bidegain and Rico (2012); Prevosti et al. (2013)
Barrancalobian	3.3–2.9 Ma	Prevosti et al. (2013)
Chapadmalalan	4.5/5–3.3 Ma	Cione et al. (2007), (2015); Woodburne (2010); Deschamps et al. (2013)
Montehermosan	5.28–4.5/5 Ma	Tomassini et al. (2013)
Huayquerian	9–5.28 Ma	Tomassini et al. (2013); Prevosti et al. (2013)
Chasicoan	10.0–9.0 Ma	compromise interval, Flynn and Swisher (1995)
Mayoan	11.8–10 Ma	compromise interval, Flynn and Swisher (1995)
Laventan	13.8–11.8 Ma	Flynn and Swisher (1995)
Colloncuran	15.5–14 Ma	Flynn and Swisher (1995)
Friasian	16.3–15.5 Ma	compromise interval, Flynn and Swisher (1995)
Santacrucian	18–16 Ma (coast) or ca. 19–14 Ma (Andean foothill)	Vizcaíno et al. (2012); Perkins et al. (2012)
Colhuehuapian	21.0–20.1	Dunn et al. (2012)
Deseadan	29.4–24.2 Ma, or possibly 30–23 Ma	Dunn et al. (2012)
Tinguiririran	33.6–31.3 Ma	Dunn et al. (2012)
Mustersan	39–36.5 Ma	Woodburne et al. (2014a)
Casamayoran	45–39.5 Ma	Vacan + Barrancan (Carlini et al. 2010; Dunn et al. 2012; Woodburne et al. 2014a)
Riochican	ca. 49 Ma	Woodburne et al. (2014b)
Itaboraian	53–50 Ma	Woodburne et al. (2014b)
Peligran	63.8–63.2 Ma	Woodburne et al. (2014b)
Tiupampan	ca. 64 Ma	Woodburne et al. (2014a), (b)

1.2 Abbreviations

1.2.1 Institutional

AMNH, American Museum of Natural History, New York, USA

BM, British Museum of Natural History, London, UK

CICYTTP-PV-M, Centro de Investigaciones Científicas y Transferencia de Tecnología a la Producción de Diamante, Laboratorio de Paleontología de Vertebrados, Diamante, Entre Ríos, Argentina

CORD-PZ, Museo de Paleontología, Facultad de Ciencias Exactas, Físicas y Naturales de la Universidad Nacional de Córdoba, Córdoba, Argentina

FMNH, Field Museum of Natural History, Chicago, USA

IGM, Instituto Nacional de Investigaciones en Geociencias, Minería y Química, Bogotá, Colombia

IVIC OR, Instituto Venezolano de Investigaciones Científicas, Orocual collection, Altos de Pipe, Miranda, Venezuela

LACM-HC, Natural History Museum of Los Angeles County, Hancock collection, Los Angeles, USA

MACN, Museo Argentino de Ciencias Naturales "Bernardino Rivadavia," Ciudad Autónoma de Buenos Aires, Argentina (MACN-A, Ameghino collection; MACN-PV, Vertebrate Paleontology collection)

MLP, Museo de La Plata, La Plata, Buenos Aires, Argentina (MLP-PV, Vertebrate Paleontology collection; MLP-M Mammalogy collection)

MMP, Museo Municipal de Mar de Plata "Lorenzo Scaglia," Mar del Plata, Buenos Aires, Argentina

MNHN-Bol, Museo Nacional de Historia Natural, La Paz, Bolivia

MPM-PV, Museo Regional Provincial "Padre M. J. Molina," Santa Cruz, Argentina

MPS, Museo Paleontológico de San Pedro, San Pedro, Buenos Aires, Argentina

NRM, Swedish Museum of Natural History, Sweden

UCMP, University of California Museum of Paleontology, California, USA

VF, Museo Royo y Gómez, Universidad Central de Venezuela, Caracas, Venezuela

YPM PU, Yale Peabody Museum, Princeton University collection, New Haven, USA

1.2.2 Anatomical

Abbreviations for the dentition are: I/i, incisor, C/c, canine, P/p, premolar, and M/m, molar; upper and lower case letters refer to upper and lower teeth, respectively; d indicates deciduous dentition.

1.2.3   Others Abbreviations and Conventions

BM, Body mass
FAD, First Appearance Datum
GABI, Great American Biotic Interchange
ka, *kiloannum,* unit of time equal to one thousand years
kg, kilograms
Ma, *megannum,* unit of time equal to one million years on the radioisotopic
timescale
masl, meters above sea level
NA, North America
RGA, relative grinding area index
SA, South America
WN, without collection number.

1.3   Nomenclature

1.3.1   Biological Conventions

Mammalia refers to the most recent common ancestor of the living monotremes,
marsupials and placentals, and all its descendents (Rowe 1988; Fig. 1.5). Eutheria
is the group that includes placentals and all the taxa more closely related to placentals than to marsupials, and similarly Metatheria includes marsupials and all the
taxa more closely related to marsupials than to placentals (Fig. 1.5). Placentalia is a
crown group and includes the last common ancestor of extant placental species
(humans, horses, dogs) and all its descendents. The counterpart Marsupialia is the
crown group that includes the last common ancestor of extant marsupial species
(opossums, kangaroos, koalas) and all its descendents (Rougier et al. 1998;
Fig. 1.5).

In our text, the terms "carnivoran" refers to any member of the taxonomic order
Carnivora while "carnivore" refers to any flesh-eating taxa (regardless of taxonomic
identity).

1.3.2   Ecological Conventions

The term "diversity" refers to the taxonomic richness (measured as the total number
of species) during a given period of time. The term "disparity" refers to morphological heterogeneity within a species or clade (e.g., Foote 1997; Hughes et al.
2013). We mean by "ecological disparity" the area in the morphospace defined by

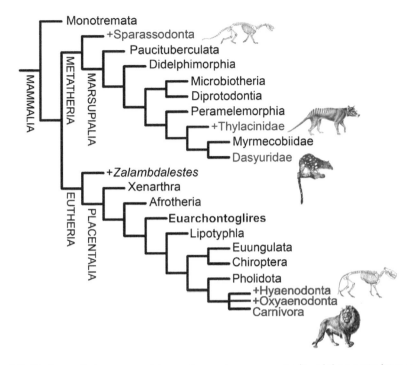

Fig. 1.5 Phylogenetic tree summarizing major carnivore groups (red) and the nomenclature used in the text

ecological variables represented by a taxonomic group or fauna. "Adaptive radiation" implies both, a proliferation in numbers of taxa and a diversification of form (Simpson 1953; Foote 1997). In our analyses, taxonomic richness includes the known record as well as any "range-through taxon" (="Lazarus taxon"; i.e., one taxon found in underlying and overlying ages or strata, but absent in the intermediate ones; Smith 1994), by assuming that the absences are observational artifacts. The time units for the analyses are the South American Ages/Stages (Table 1.1).

Following Van Valen (1971:421), "adaptive zone" refers to the niche of any taxon including a supraespecific one. We assumed that two taxa occupied a similar adaptive zone if they had similar ecological requirements (e.g., body size and diet).

The change through time of a given adaptive zone is explained by two processes: "competitive displacement" and "opportunistic replacement" (e.g., Benton 1983; Krause 1986). Competitive displacement implies the existence of temporal, geographic, and ecological overlap of two clades occupying the same adaptive zone under two conditions: (1) the diversity of one clade declines while the diversity of a second clade increases; (2) these changes in diversity are not associated with climatic or floral changes. Alternatively, opportunistic replacement implies that (1) one clade radiates coincident upon the extinction of the other; (2) the rate of replacement is rapid; (3) the two clades are not found together or the replaced group is found when the other is not dominant; (4) the replacement could be associated with climatic or floral changes (Benton 1983; Krause 1986).

Following Prevosti et al. (2013), body size was arbitrarily divided into three categories according to mass: small (below 7 kg), medium (between 7 and 15 kg), and large (above 15 kg). Diet was classified as hypercarnivorous (species that feed mostly on other vertebrates chiefly mammals), mesocarnivorous (species with diets mostly composed of vertebrates but with important consumption of insects, fruits, or other non-vertebrate items), and omnivorous (species that incorporate a large proportion of non-vertebrate items, such as insects or vegetables) (cf. Van Valkenburgh and Koepfli 1993). The diet was inferred by dental morphology using a index based on the relative grinding area (RGA; modified from Van Valkenburgh 1991) of the most carnivorous molar (m4 for Sparassodonta and m1 for Carnivora) (Prevosti et al. 2013). In this context, a taxon was considered hypercarnivorous, when the RGA was lower than 0.48; mesocarnivorous when the index was between 0.48 and 0.54; and omnivorous when it was larger than 0.54.

References

Argot C (2004) Evolution of South American mammalian predators (Borhyaenoidea): anatomical and palaeobiological implications. Zool J Linn Soc 140:487–521

Babot J, Forasiepi AM (2016) Mamíferos predadores nativos del Cenozoico sudamericano: evidencias filogenéticas y paleoecológicas. In: Agnolin FA, Lio GL, Brissón Egli F, Chimento NR, Novas FE (eds) Historia Evolutiva y Paleobiogeográfica de los Vertebrados de América del Sur. Contribuciones MACN, vol 6. Museo Argentino de Ciencias Naturales "Bernardino Rivadavia", Buenos Aires, pp 219–230

Benton MJ (1983) Dinosaur success in the Triassic: a noncompetitive ecological model. Q Rev Biol 58:29–55

Bidegain JC, Rico Y (2012) Magnetostratigraphy and magnetic parameters of a sedimentary sequence in Punta San Andres, Buenos Aires, Argentina. Quat Int 253:91–103. doi:10.1016/j.quaint.2011.08.018

Biknevicius AR, Van Valkenburgh B (1996) Design for killing: craniodental adaptations of predators. In: Gittleman JL (ed) Carnivore behavior, ecology, and evolution, vol 2. Cornell University Press, New York, pp 393–428

Carlini AA, Ciancio M, Scillato-Yané GJ (2010) Middle Eocene—early Miocene Dasypodidae (Xenarthra) of southern South America: faunal succession at Gran Barranca—biostratigraphy and paleoecology. In: Madden RH, Carlini AA, Vucetich MG, Kay RF (eds) The paleontology of Gran Barranca. Cambridge University Press, Cambridge, pp 106–129

Cione AJ, Gasparini GM, Soibelzon E, Soibelzon LH, Tonni EP (2015) The Great American Biotic Interchange: a South American perspective. Springer Earth System Sciences, Dordrecht

Cione AL, Tonni EP et al (2007) Mamíferos continentales del Mioceno tardío a la actualidad en la Argentina:cincuenta años de estudios. Publicación Especial Ameghiniana 50° aniversario. APA, Buenos Aires, pp 257–278

Crusanfont-Pairó M, Truyols-Santoja J (1956) A biometric study of the evolution of fissiped carnivores. Evolution 10:314–332

Deschamps CM, Vucetich MG, Montalvo CI, Zárate MA (2013) Capybaras (Rodentia, Hydrochoeridae, Hydrochoerinae) and their bearing in the calibration of the late Miocene-Pliocene sequences of South America. J South Am Earth Sci 48:145–158

Dunn RE, Madden RH, Kohn MJ, Schmitz MD, Strömberg CAE, Carlini AA, Re GH, Crowley J (2012) A new chronology for middle Eocene—Early Miocene South American Land Mammal Ages. Geol Soc Am Bull 125:539–555

Ercoli MD, Prevosti FJ (2012) Estimación de masa de las especies de Sparassodonta (Mammalia, Metatheria) de edad Santacrucense (Mioceno temprano) a partir del tamaño del centroide de los elementos apendiculares: inferencias paleoecológicas. Ameghiniana 48:462–479

Ercoli MD, Prevosti FJ, Forasiepi AM (2014) The structure of the mammalian predator guild in the Santa Cruz Formation (late early Miocene). J Mammal Evol 21:369–381

Evans AR, Fortelius M (2008) Three-dimensional reconstruction of tooth relationships during carnivoran chewing. Palaeontologia Electronica 11(2):1–11

Ewer RF (1973) The carnivores. Cornell University Press, New York

Figueirido B, Janis CM (2011) The predatory behaviour of the thylacine: Tasmanian tiger or marsupial wolf? Biol Lett 7:937–940

Figueirido B, Macleod N, Krieger J, De M, Pérez-claros JA, Palmqvist P, Pe JA (2011) Constraint and adaptation in the evolution of carnivoran skull shape. Paleobiology 37:490–518

Figueirido B, Tseng ZJ, Martín-Serra A (2013) Skull shape evolution in durophagous carnivorans. Evolution 67:1975–1993

Flynn JJ, Swisher CC III (1995). South American Land Mammal Ages: correlation to global geochronologies. In: Berggren WA (ed) Geochronology, time scales and global stratigraphic correlation. Society for Sedimentary Geology Special Publication 54, pp 317–333

Flynn JJ, Wesley-Hunt GD (2005) Carnivora. In: Rose KD, Archibald JD (eds) The rise of placental mammals. The Johns Hopkins University Press, Baltimore and London, Origins and relationships of the major clades, pp 175–198

Foote M (1997) The evolution of morphological diversity. Annu Rev Ecol Syst 28:129–152

Forasiepi AM (2009) Osteology of *Arctodictis sinclairi* (Mammalia, Metatheria, Sparassodonta) and phylogeny of Cenozoic metatherian carnivores from South America. Monografías Mus Arg Sci Nat "Bernardino Rivadavia" (n.s.) 6:1–174

Forasiepi AM, Martinelli AG, Goin FJ (2007) Revisión taxonómica de *Parahyaenodon argentinus* Ameghino y sus implicancias en el conocimiento de los grandes mamíferos carnívoros del Mio-Plioceno de América del Sur. Ameghiniana 44:143–159

Freeman PW (1984) Functional cranial analysis of large animalivorous bats (Microchiroptera). Biol J Linn Soc 21:387–408

Friscia AR, Van Valkenburgh B, Biknevicius AR (2007) An ecomorphological analysis of extant small carnivorans. J Zool 272:82–100

Goin FJ (1989) Late Cenozoic South American marsupial and placental carnivores: changes in predator-prey evolution. 5° International Theriological Congress, Abstracts: 271–272

Goin FJ, Pardiñas UFJ (1996) Revision de las especies del genero *Hyperdidelphys* Ameghino, 1904 (Mammalia, Marsupialia, Didelphidae). Su significación filogenética, estratigráfica y adaptativa en el neógeno del Cono Sur Sudamericano. Estud Geol 52:327–359

Goin FJ, Woodburne MO, Zimicz AN, Martin GM, Chornogubsky L (2016) A brief history of South American metatherians, evolutionary contexts and intercontinental dispersals Springer, Dordrecht

Hunt RM Jr (1996) Biogeography of the order carnivora. In: Gittleman JL (ed) Carnivore behavior, ecology, and evolution, vol 2. Cornell University Press, New York, pp 485–541

Hughes M, Gerber S, Wills MA (2013) Clades reach highest morphological disparity early in their evolution. Proc Natl Acad Sci 110:13875–13879

Jones ME (2003) Convergence in ecomorphology and guild structure among marsupial and placental carnivores. In: Jones ME, Dickman C, Archer M (eds) Predators with pouches: the biology of carnivorous marsupials. CSIRO Publications, Melbourne, pp 285–296

Jones ME, Stoddart DM (1998) Reconstruction of the predatory behaviour of the extinct marsupial thylacine (*Thylacinus cynocephalus*). J Zool 246:239–246

Krause DW (1986) Competitive exclusion and taxonomic displacement in the fossil record: the case of rodents and multituberculates in North America. In: Flanagan KM, Lillegraven JA (eds) Vertebrates, phylogeny and philosophy. University of Wyoming Contributions to Geology, Special Paper 3, pp 119–130

Lewis M, Morlo M (2010) Creodonta. In: Werdelin L (ed) Cenozoic mammals of Africa. University of California Press, Berkeley, pp 549–566

Marshall LG (1978) Evolution of the Borhyaenidae, extinct South American predaceous marsupials. Univ Calif Press Geol Sci 117:1–89

Martin LD (1989) Fossil history of the terrestrial Carnivora. In: Gittleman JL (ed) Carnivore behavior, ecology, and evolution, vol 1. Cornell University Press, New York, pp 536–568

Meloro C, O'Higgins P (2011) Ecological adaptations of mandibular form in fissiped Carnivora. J Mammal Evol 18:185–200

Nowak RM (2005) Walker's carnivores of the world. The John Hopkins University Press, Baltimore

Patterson B, Pascual R (1972) The fossil mammal fauna of South America. In: Keast A, Erk FC, Glass B (eds) Evolution, mammals, and Southern continents. State University of New York Press, Albany, pp 247–309

Peigné S, Chaimanee Y, Yamee C, Marandat B, Srisuk P, Jaeger J-J (2009) An astonishing example of convergent evolution toward carnivory: Siamosorex debonisi n. gen., n. sp. (Mammalia, Lipotyphla, Soricomorpha, Plesiosoricidae) from the latest Oligocene of Thailand. Geodiversitas 31:973–992

Perkins ME, Fleagle JG, Heizler MT, Nash B, Bown TM, Tauber AA, Dozo MT (2012). Tephrochronology of the Miocene Santa Cruz and Pinturas formations, Argentina. In: Vizcaíno SF, Kay RF, Bargo MS (eds) Early Miocene paleobiology in Patagonia: high-latitude paleocommunities of the Santa Cruz Formation, Cambridge University Press, Cambridge, pp 23–40

Prevosti FJ, Soibelzon LH (2012) The evolution of South American carnivore fauna: a paleontological perspective. In: Patterson B, Costa LP (eds) Bones, clones and biomes: the history and geography of recent neotropical mammals. University Chicago Press, Chicago, pp 102–122

Prevosti FJ, Turazzini GF, Ercoli MD, Hingst-Zaher E (2012a) Mandible shape in marsupial and placental carnivorous mammals: morphological comparative study using geometric morphometry. Zool J Linn Soc 164:836–855

Prevosti JP, Forasiepi AM, Ercoli MD, Turazzini GF (2012b) Paleoecology of the mammalian carnivores (Metatheria, Sparassodonta) of the Santa Cruz Formation (late early Miocene). In: Vizcaíno SF, Kay RF, Bargo MS (eds) Early Miocene paleobiology in Patagonia: high-latitude paleocommunities of the Santa Cruz formation. Cambridge University Press, Cambridge, pp 173–193

Prevosti FJ, Forasiepi A, Zimicz N (2013) The evolution of the Cenozoic terrestrial mammalian predator guild in South America: competition or replacement? J Mammal Evol 20:3–21

Rasmussen DT, Gutierrez M (2009) A mammalian fauna from the late Oligocene of northwestern Kenya. Palaeontogr Abt A 288:1–52

Rodrigo L. Tomassini, Claudia I. Montalvo, Cecilia M. Deschamps, Teresa Manera, (2013) Biostratigraphy and biochronology of the Monte Hermoso Formation (early Pliocene) at its type locality, Buenos Aires Province, Argentina. Journal of South American Earth Sciences 48:31–42

Rougier GW, Wible JR, Novacek MJ (1998) Implications of Deltatheridium specimens for early marsupial history. Nature 396:459–463

Rowe T (1988) Definition, diagnosis, and origin of Mammalia. J Vert Paleontol 8:241–264

Savage RJG (1977) Evolution in carnivorous mammals. Palaeontology 20:237–271

Simpson GG (1950) History of the fauna of Latin America. Am Sci 38:361–389

Simpson GG (1953) The major features of evolution. Columbia University Press, New York

Simpson GG (1969) South American mammals. In: Fitkau EJ, Illies J, Klinge H, Schwabe GH, Sioli H (eds) Biogeography and ecology in South America. Dr. W Junk Publishers, The Hague, pp 876–909

Simpson GG (1971) The evolution of marsupials in South America. An Acad Bras Cs 43:103–118

Simpson GG (1980) Splendid isolation. In: The curious history of South American mammals. Yale University Press, New Haven

Smith AB (1994) Systematics and the fossil record: documenting evolutionary patterns. Blackwell Scientific Publications, Oxford

Solé F, Smith T (2013) Dispersals of placental carnivorous mammals (Carnivoramorpha, Oxyaenodonta & Hyaenodontida) near the Paleocene-Eocene boundary: a climatic and almost worldwide story. Geologica Belgica 16(4):254–261

Strahan R (1995) The mammals of Australia. Reed Books, Sydney

Tomassini RL, Montalvo CI, Deschamps C, Manera T (2013) Biostratigraphy and biochronology of the Monte Hermoso Formation (Early Pliocene) at its type locality, Buenos Aires Province, Argentina. J South Am Earth Sci 48:31–42

Van Valen L (1971) Adaptive zones and the orders of mammals. Evolution 25:420–428

Van Valkenburgh B (1989) Carnivore dental adaptation and diet: a study of trophic diversity within guilds. In: Gittleman JL (ed) Carnivore behavior, ecology, and evolution, vol 1. Cornell University Press, New York, pp 410–436

Van Valkenburgh B (1991) Iterative evolution of hypercarnivory in canids (Mammalia: Carnivore): evolutionary interactions among sympatric predators. Paleobiology 17:340–362

Van Valkenburgh B (1999) Major patterns in the history of carnivorous mammals. Annu Rev Earth Planet Sci 27:463–493

Van Valkenburgh B, Koepfli KP (1993) Cranial and dental adaptation to predation in canids. Symp Zool Soc Lond 65:15–37

Vizcaíno SF, Kay RF, Bargo MS (2012) Background for a paleoecological study of the Santa Cruz Formation (late early Miocene) on the Atlantic coast of Patagonia. In: Vizcaíno SF, Kay RF, Bargo MS (eds) Early miocene paleobiology in Patagonia: high-latitude paleocommunities of the Santa Cruz Formation. Cambridge University Press, Cambridge, pp 1–22

Wang X, Tedford R, Antón M (2008) Dogs:their fossil relatives and evolutionary history. Columbia University Press, Chichester

Webb SD (1985) Late Cenozoic mammal dispersals between the Americas. In: Stheli FG, Webb SD (eds) The Great American Biotic Interchange. Plenum Press, New York, pp 357–386

Werdelin L (1987) Jaw geometry and molar morphology in marsupial carnivores: analysis of a constraint and its macroevolutionary consequences. Paleobiology 13:342–350

Werdelin L (1996) Carnivoran ecomorphology: a phylogenetic perspective. In: Gittleman JL (ed) Carnivore behavior, ecology, and evolution, vol 2. Cornell University Press, New York, pp 582–624

Werdelin L (2010) Cenozoic mammals of Africa. University of California Press, Berkeley, pp 1–960

Wilson DE, Mittermeier RA (2009) Handbook of the mammals of the world, vol 1 Carnivores. Lynx Editions, Barcelona

Wilson DE, Reeder DM (2005) Mammal species of the world: a taxonomic and geographic reference, 3rd edn. Johns Hopkins University Press, Baltimore

Woodburne MO (2010) The Great American Biotic Interchange: dispersals, tectonics, climate, sea level and holding pens. J Mammal Evol 17:245–264

Woodburne M, Cione AL, Tonni EP (2006) Central American provincialism and the Great American Biotic Interchange. In: Carranza-Castañeda O, Lindsay EH (eds) Advances in late Tertiary vertebrate paleontology in Mexico and the Great American Biotic Interchange. Publicación Especial del Instituto de Geología y Centro de Geociencias de la Universidad Nacional Autónoma de México, vol 4, pp 73–101

Woodburne MO, Goin FJ, Bond M, Carlini AA, Gelfo JN, López GM, Iglesias A et al (2014a) Paleogene land mammal faunas of South America; a response to global climatic changes and indigenous floral diversity. J Mammal Evol 21:1–73

Woodburne MO, Goin FJ, Raigemborn MS, Heizler M, Gelfo JN, Oliveira EV (2014b) Revised timing of the South American early Paleogene land mammal ages. J South Am Earth Sci 54:109–119

Wroe S (2002) A review of terrestrial mammalian and reptilian carnivore ecology in Australian fossil faunas, and factors influencing their diversity: the myth of reptilian domination and its broader ramifications. Aust J Zool 50:1–24

Wroe S, Milne N (2007) Convergence and remarkably consistent constraint in the evolution of carnivore skull shape. Evolution 61:1251–1260

Wroe S, Brammall J, Cooke B (1998) The skull of *Ekaltadeta ima* (Marsupialia, Hypsiprymnodontidae) an analysis of some marsupial cranial features and a reinvestigation of propleopine phylogeny, with notes on the inference of carnivory in mammals. J Paleontol 72:735–751

Wroe S, Myers T, Seebacher F, Kear B, Gillespie A, Crowther M, Salisbury S (2003) An alternative method for predicting body-mass: the case of the marsupial lion. Paleobiology 29:404–412

Wroe S, Clausen P, McHenry C, Moreno K, Cunningham E (2007) Computer simulation of feeding behaviour in the thylacine and dingo as a novel test for convergence and niche overlap. Proc R Soc Biol 274:2819–2828

Zimicz N (2014) Avoiding competition: the ecological history of late Cenozoic metatherian carnivores in South America. J Mammal Evol 21:383–393

Chapter 2
Paleoenvironment, Tectonics, and Paleobiogeography

Abstract The Earth experienced dramatic transformations during the Cenozoic, with changing sea levels, climate, and tectonic events having major influences on the global biota. In South America, loss of the connection between Patagonia and Antarctica, Andean orogeny, and formation of the Isthmus of Panama defined the continent, as we know it today. These events had enormous effects on local faunas, with major consequences for their evolution and extinction. The Great American Biotic Interchange (GABI), a major natural experiment in biotic reorganization, was either enabled or at least greatly enhanced by land connections between North and South America during the late Neogene. The outcome of the meeting of previously separated biotas was a drastic change, both for the composition of South American faunas and the terrestrial ecosystems they inhabited.

Keywords Paleogeographic changes · Marine transgressions · Montane tectonics Orogenic rain shadows · Glacial cycles

2.1 Historical Factors: An Overview

The faunal and floral communities of South America experienced strong climatic, tectonic, and geographic changes during the Cenozoic. These historical causal factors profoundly affected the evolution of South American terrestrial ecosystems during the past 66 Ma. In this chapter, we review key aspects of South American geological history during the Cenozoic, focusing on those features that likely had the greatest influence on the environment and fauna. This chapter is a summary of major patterns of tectonic, climatic, geographic, environmental, and faunal change as they occurred in South America. As part of our scope, we include evaluations of controversies concerning the timing of the establishment of the Panamanian land bridge and the chronology of the GABI.

© Springer International Publishing AG 2018
F.J. Prevosti and A.M. Forasiepi, *Evolution of South American Mammalian Predators During the Cenozoic: Paleobiogeographic and Paleoenvironmental Contingencies*, Springer Geology, https://doi.org/10.1007/978-3-319-03701-1_2

2.1.1   Geography and Climate—from Greenhouse to Icehouse

At the beginning of the Cenozoic, South America was separated from North America by a major seaway, but still connected to Antarctica and (at a further remove) Australia. Temperatures were warm worldwide, a persistence of greenhouse conditions from the close of the Mesozoic. The terminal late Paleocene–early Eocene ($\sim$59 to $\sim$52 Ma) was a period of unusual warmth even by the standards of previous eras. Global temperatures and atmospheric greenhouse gas levels were higher than they have been at any other point in the past 65 Ma (Fig. 2.1). This hyperthermal period, known as the Paleocene–Eocene thermal maximum (PETM), peaked at $\sim$52–50 Ma, with the early Eocene climatic optimum (EECO) (Woodburne et al. 2014a, b).

By the early Eocene, global surface ocean currents were warm, with saline concentrations similar to those at depth. Equatorial waters circulated with those of the South Pacific and South Atlantic, keeping ocean temperatures uniform between high and low latitudes (Fig. 2.2).

Fig. 2.1 Changes in atmospheric CO_2 levels and global climate during the Cenozoic. Abbreviations: EECO, early Eocene climatic optimum; MMCO, middle Miocene climatic optimum; PETM, Paleocene–Eocene thermal maximum (based on Zachos et al. 2001)

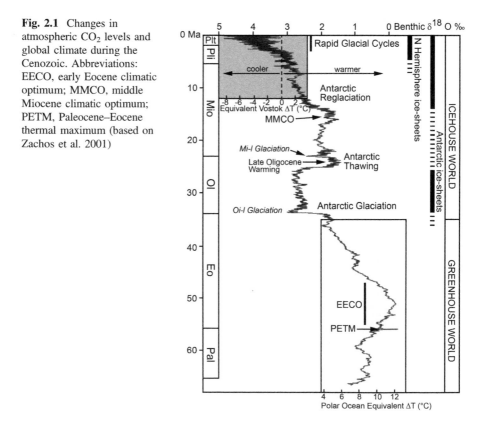

Fig. 2.2 Ocean circulation pattern in the Southern Hemisphere at the early Eocene and early Oligocene. The Antarctic Circumpolar Current (CCA, in blue) is a clockwise cold ocean current defined after the separation of Antarctic, Australia, and South America. (based on Benedetto 2010)

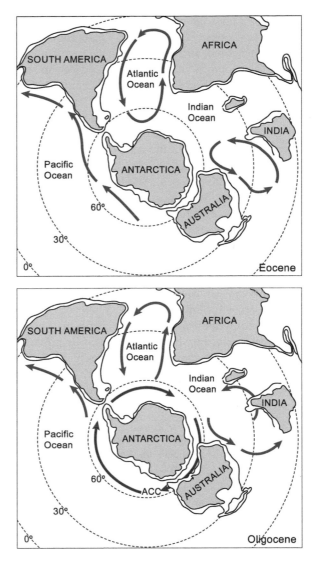

The EECO was followed by a long trend toward cooler conditions from 50 to 34 Ma. This cooling event peaked at the Eocene–Oligocene Boundary (~34 Ma), and the world then began to change from greenhouse to icehouse (Fig. 2.1). Ice sheets previously present at higher altitudes in the Antarctic expanded, ushering in the first major Cenozoic glaciation on that continent (Oi-1 Glaciation). Temperatures declined 7 °C in deep seas (from ~12 to ~4.5 °C) (Zachos et al. 2001; Reguero et al. 2013), and significant temperature gradients appeared from the equator to the poles.

The sea floor between Antarctica and Australia started forming by the middle/ early late Eocene. By the early Oligocene (~ 32 Ma), ocean depth permitted deep-water communication between the Pacific and Indian Oceans, setting the stage for the development of a major Antarctic ice cap and the largest Cenozoic sea level drop (~ 30 Ma; Haq et al. 1987). The current oceanic configuration was complemented by the opening of the Drake Passage. The precise timing of the separation of Antarctica and South America is difficult to establish. The evidence indicates that by the late Oligocene (~ 24 Ma; Pfuhl and McCave 2005), the deep-water connection between the Pacific and Atlantic Oceans was fully established. However, studies based on seafloor magnetic anomalies indicated that a shallow-water opening (<1000 m deep) was present as early as ~ 50 Ma (early Eocene; Livermore et al. 2007; see also Scher and Martin 2006; Reguero et al. 2013). This configuration resulted in the formation of the Antarctic Circumpolar Current (ACC), a clockwise current flowing around Antarctica (Fig. 2.2) that thermally isolated the continent for the remainder of the Cenozoic and profoundly influenced worldwide climate.

The Eocene–Oligocene boundary is marked by biotic changes on a global basis, including major diversification/extinction and immigration/emigration events (Prothero 1994; Sluijs et al. 2007; Wing et al. 1995). In Europe, this episode was recognized by the Swiss paleontologist Hans Stehlin as "La Grande Coupure," or the great break, in what was perceived as previous long-term faunal continuity. Similar faunal reorganizations were later recognized for North America by Henry F. Osborn (Prothero 1994), Asia with the concept of the "Mongolian Remodeling" (Meng and McKenna 1998), and South America with the "Patagonian Hinge," based on an abrupt change within metatherian-dominated associations in Patagonia (Goin et al. 2010, 2016).

Climatic deterioration at the start of the Oligocene and the formation of semipermanent ice sheets in Antarctica persisted until the late Oligocene (26– 27 Ma; Zachos et al. 2001), when a warming trend reduced Antarctic ice cover. Another hypothermal peak is recorded in the early Miocene (Mi-1 Glaciation). From this point onward until the middle Miocene, warmer conditions prevailed. Global ice volume remained low, and bottom water temperatures trended slightly higher. This warm phase peaked in the middle Miocene (~ 17 to ~ 15 Ma) with the middle Miocene climatic optimum (MMCO) (Fig. 2.1). Global temperatures gradually decreased once more, with the initiation of the modern phase of Antarctic glaciation, reestablishing permanent ice cover by 10 Ma (Zachos et al. 2001). Global temperatures decreased gradually through the late Miocene until the early Pliocene (6 Ma), with additional cooling and ice sheet expansion in West Antarctica and the Arctic.

The early Pliocene is marked by a subtle warming trend until ~ 3.2 Ma when temperatures abruptly decreased, marking the onset of Northern Hemisphere Glaciation (NHG). Since then, temperature curves show contrasting peaks and valleys, correlated with the pattern of repeated glaciations and interglaciations that characterize the Ice Ages of the last 2.6 Ma. The amplitude of these changes greatly increased at the beginning of the Middle Pleistocene 0.7 Ma (Zachos et al. 2001).

This event has been previously correlated with the completion of the Isthmus of Panama (but see below) and consequent changes in oceanic circulation (Fig. 2.3). Closure of the seaway between the Pacific and Atlantic Oceans enhanced the Gulf Stream, the warm Atlantic current that passes from the Gulf of Mexico along the North American Atlantic seaboard before departing for northern Europe. Enhanced production of water vapor from evaporation of the warm waters of the Gulf Stream triggered increased snowfall in Europe, Asia, and North America and the consequent formation of thick snow cover (Haug and Tiedemann 1998).

In Patagonia, glaciations had begun to occur in the southern Andes (Santa Cruz Province) by the late Miocene–early Pliocene ∼7–4.4 Ma (Rutter et al. 2012). After this, 15 successive expansion/retreat events have been documented for southern South America, some of which can be recognized worldwide (Rutter et al. 2012). During the late Pliocene (4.8–4.7 Ma, 3.5–2.79 Ma), extra-Andean ice sheets formed in Patagonia, with major expansions (the so-called Great Patagonian Glaciation) dating to ∼2.4 Ma and ∼1 Ma (Rabassa 2008; Rutter et al. 2012). Evidence of early Pleistocene expansions (∼2.6 Ma) has been recovered in the Andes of Bolivia and Colombia, while that for the mid-Pleistocene glaciation (∼1 Ma) has been recovered from the Bogota basin of Colombia (Rutter et al. 2012). The Last Glacial Maximum (LGM) in the Northern Hemisphere is recorded

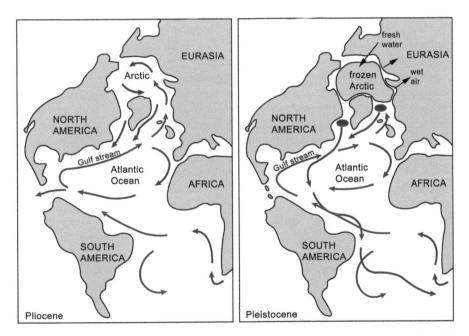

Fig. 2.3 Ocean circulation pattern before and after the closure of the Panama Isthmus. The Gulf Stream is a warm Atlantic Ocean current from the Gulf of Mexico to northern Europe enhanced by the closure of Panama Isthmus. The water vapor from evaporation of the warm waters of the Gulf Stream triggered increased snowfall in Europe, Asia, and North America and the consequent formation of thick snow cover (based on Benedetto 2010)

from ~80 to 20 Ka (Wisconsinan glaciation), while in South America it ranges from ~48 to 25 Ka (Rutter et al. 2012). In South America, evidence for this later advance is restricted to the cordilleras and valleys of the Andes (Clapperton 1993; Rabassa 2008; Rutter et al. 2012). Ice sheets reached their maximum ~26,500 years ago, with deglaciation starting ~19,000 years ago (Clark et al. 2009), ushering in the current interglacial period from about 11,700 years ago.

2.1.2 Tectonics—Rise of the Andes

The Andes, originating by subduction of oceanic crust underneath the Pacific margin of South America, constitutes the world's longest active orogenic system (Ramos 2009). Its range extends more than 8000 km, from the Caribbean Sea in the north to Tierra del Fuego in the south, with an average height of about 4000 masl and maximum elevations up to 7000 masl (Ramos 1999).

Subduction, initiated in the late Proterozoic, has been episodic up to the present (Ramos 2009). In southern South America, several phases of Cenozoic Andean orogeny are denoted as the Incaica (~30 Ma), Pehuenche (~25 Ma), Quechua (~15.5 Ma), and Diaguita Phases (~4.5 Ma) (Yrigoyen 1979; Leanza and Hugo 1997). These events, defined at the latitudinal level of northern Patagonia (Neuquén Province), are diachronic along the major segments, usually identified as the northern, central, and southern Andes (Fig. 2.4). Each segment has its own history with regard to tectonics, including convergence rate of oceanic and continental plates, collision with allochthonous terrains, subduction angle, age and rigidity of oceanic crust, magmatism, and sedimentation. Normal subduction (angle about 30°) is characterized by intense vulcanism. Flat-slab subduction, with inclination between 5° and 10° (subhorizontal), is characterized by significant seismic activity, minor volcanism, and substantial compression forces (Ramos 1999) (Fig. 2.4).

With regard to diachronicity, the Principal Cordillera (about 33°S) was raised between 20 and 8.6 Ma, the Frontal Cordillera between 10 and 3 Ma, and the Precordillera after 4 Ma (Ramos 1999). The Cordillera Oriental was fully uplifted ~8 Ma in the extreme northwest of Argentina (Jujuy and Salta Provinces), while the Sierras Subandinas did not achieve maximum uplift until the early Pliocene (Ramos 1999). The level of uplift was such that the northwestern Argentinean Andes reached altitudes greater than 6000 m prior to 6 Ma.

The central Andes (the Altiplano of Bolivia, northern Chile, and southern Peru) underwent a steep uplift starting ~9 Ma, with the northern part reaching modern altitudes by ~8 Ma and the south by 3.6 Ma (Garzione et al. 2008; Bershaw et al. 2010).

In northern central and northern Andes (north of 20°S), present-day elevations were achieved during or soon after 10–8 Ma. Andean orogeny in Ecuador, Colombia, and Venezuela included collision with island arcs and oceanic plateaus (Iturralde-Vinent and MacPhee 1999; Coates et al. 2004; Ramos 2009; Farris et al. 2011; Montes et al. 2012; Coates and Stallard 2013). The Venezuelan Andes

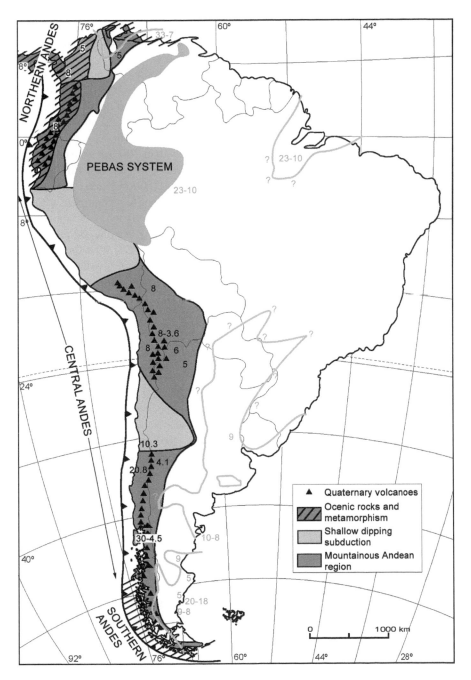

Fig. 2.4 Major segments of the Andes, Quaternary volcanic zones and tectonic processes, and in the blue lines, major Neogene marine ingressions (adapted from Ramos 1999; Hoorn et al. 2010; Del Río et al. 2013). Dating of the events are expressed in Ma

experienced rapid uplift and widening during the late Miocene (Erikson et al. 2012), with uplift in Merida, Sierra de Perija, Santa Marta, Lara-Falcón, and Guajira occurring ∼5 Ma.

In Colombia, the rise of the Eastern Cordillera was complex, involving several steps. Depending on the area, uplift occurred at intervals between the Oligocene and late Miocene (∼30–26 Ma, ∼23–20 Ma, and ∼10–6 Ma) and involved an overall rise of more than 1000 m (Ochoa et al. 2012). The Western and Central Cordilleras had emerged by the Paleocene–Eocene, with rejuvenation in the Miocene (Borrero et al. 2012). The Bogota basin in Colombia was less than 1000 m above sea level until the early Pliocene, when it rose to 1500 m. Between the late Pliocene and early Pleistocene, it reached its modern height of about 2500 m (Rutter et al. 2012).

For obvious reasons, the Andes constitutes an important biogeographic structure in South America. Its physiographic structure, in combination with the drastic environmental changes occasioned by their uplift, has long acted as biogeographic barriers for some taxa and corridors for others (Webb 1985, 1991; Woodburne 2010). In consequence, they have, in effect, ruled the later Cenozoic evolution of most terrestrial ecosystems in South America (Hoorn et al. 2010; Patterson et al. 2012).

2.1.3   Sea Level Changes and Marine Ingressions

South America experienced several marine ingressions between the late Miocene and Pleistocene. Marine deposits dating between 10 and 9 Ma in Patagonia and to ∼9 Ma in the Pampean Region (Entre Rios Province) establish the occurrence of a late Miocene marine ingression in the southern cone (Del Río et al. 2013; Pérez 2013; Marengo 2015; for older transgressions, see Guerstein et al. 2010; Tambussi and Degrange 2013; Woodburne et al. 2014a). Evidence of another more recent marine ingression in Patagonia date to ∼5.1 Ma (Del Río et al. 2013).

Several authors have supported the contention that the late Miocene ("Paranaense" and "Entrerriense") ingression in the southern part of the continent communicated with coeval incursions in the north, resulting in a shallow epicontinental sea covering extensive portions of lowland South America (Räsänen et al. 1995; Ramos 1999; Hovikoski et al. 2007; Uba et al. 2009). The evidence for such widespread inundation is, however, either inconclusive (Hoorn et al. 2010; Marengo 2015) or contradictory (Latrubesse et al. 2010; Gross et al. 2011). By contrast, other authors have maintained that during the late Miocene, a giant lake, the Pebas system, occupied western Amazonia (Fig. 2.4). About 10 Ma this was replaced by a large fluvial system that connected several isolated basins (e.g., Amazon and Parana). In the Pliocene, the Amazonian region was elevated and the current river-systems became entrenched (Campbell et al. 2006; Cozzuol 2006; Hoorn et al. 2010).

Pleistocene glaciations produced recurrent changes in sea level (above the present level in interglacial periods, below in glacial stages). Rises caused brief marine ingressions, mainly limited to coastal areas and river basins (Clapperton 1993; Ponce et al. 2011; Rabassa and Ponce 2013). On the east coast of southern South

America, low sea levels during glacial periods exposed a large portion of the continental platform, considerably increasing the extent of environments available for the terrestrial fauna (Ponce et al. 2011; Rabassa and Ponce 2013) (Fig. 2.5).

Fig. 2.5 Changes of the Atlantic coast of Pampa and Patagonia from the LGM to the middle Holocene (based on Ponce et al. 2011). The numbers correspond to different terraces or levels present at the continental shelf, indicating the sea level position. I: −140 m, corresponding to the Last Glacial Maximum (LGM) or an earlier glaciation; II: −120 m, equivalent to the LGM; III: −90 m, corresponding with 15 ka (calibrated); IV: −30 m, around 11–10 ka (calibrated)

In the Pleistocene, glacial–interglacial climatic cycling strongly affected vegetation coverage (Fig. 2.6). During colder glacial phases, more arid conditions were present, with an overall retreat of forests and their replacement by open environments (Clapperton 1993; Cione et al. 2003, 2008, 2015; Woodburne et al. 2014a; and bibliography cited there). On the other hand, the warmer interglacial phases recorded conditions more similar to those of the present (Clapperton 1993; Cione et al. 2008, 2015).

These environmental and floristic changes from the middle Miocene–Pliocene onward had notable impacts on several autochthonous groups of mammals (e.g., Pascual and Ortiz Jaureguizar 1990). Although several groups of SANU (e.g., Astrapotheria, Leontinidae, Adianthidae, Notohippidae) became completely extinct in the middle Miocene, some xenarthran taxa radiated extensively (Megalonychidae, Megatheriidae, and Mylodontidae) (Marshall and Cifelli 1990). A few lineages of SANU continued and even produced restricted radiations (e.g., Toxodontidae), but their later history was one of continuing decline during the middle Miocene–Pleistocene, followed by further steep reductions in the Pliocene and finally their complete disappearance with the Pleistocene–Holocene transition and associated megafaunal extinctions (Marshall and Cifelli 1990; Bond et al. 1995).

Within this broad interval, a marked faunal turnover occurred during the mid-Pliocene (e.g., Kraglievich 1952; Tonni et al. 1992; Vizcaíno et al. 2004). Turnover has been attributed to environmental changes resulting from Andean orogeny (e.g., Ortiz Jaureguizar et al. 1995; Cione and Tonni 2001) as well as to the catastrophic effects of a bolide that apparently struck somewhere in or near the Atlantic coast ∼ 3.3 Ma (Schultz et al. 1998; Vizcaíno et al. 2004). Another cause of particular interest here is the arrival of Holarctic invaders, to which discussion will now turn.

2.3   The Panamanian Isthmus and the "Great American Biotic Interchange"

Tectonism during the Miocene–Pliocene finally resulted in the establishment of a continuous terrestrial bridge between Central and South America (Coates et al. 2004; Farris et al. 2011) and triggered one of the most important biogeographic events of the Cenozoic: the "Great American Biotic Interchange." The biotic exchange between the Americas is typically portrayed as principally affecting mammals, birds, and turtles, but of course other vertebrates, invertebrates, and plants were also involved (Stehli and Webb 1985; Woodburne 2010; Tambussi and Degrange 2013; Bagley and Johnson 2014; De la Fuente et al. 2014; Gutiérrez-García and Vázquez-Domínguez 2013; Leigh et al. 2014). Nor were the effects of closure of the seaway limited to the land; disconnection between the Caribbean Sea and the Pacific Ocean had wide repercussions for marine biotas as well (Leigh et al. 2014).

Subduction of the Cocos and Nazca Plates under the Caribbean large igneous province produced a volcanic arc that collided with northwestern South America ∼ 12 Ma, reducing the distance between Central and South America, and creating an interstitial island chain (Coates et al. 2004; Fig. 2.7). Recent evidence suggests that the collision between Central and South America started around 25–23 Ma. By this time, Central America was a continuous peninsula, leaving only a narrow, shallow strait between the continents in the Miocene (Farris et al. 2011; Montes et al. 2012). Detrital zircons recovered from the northern Andes (Colombia) in northwestern South America have been interpreted as evidence of a terrestrial connection between

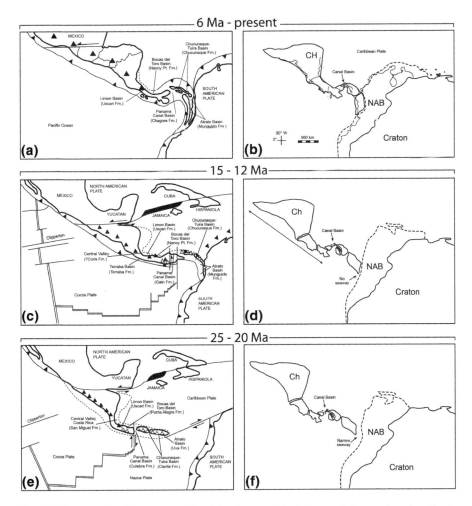

Fig. 2.7 Comparative diagrams models of the closure of the Isthmus of Panama based on Coates and Stallard (2013) for 6 Ma–present (**a, b**), 15–12 Ma (**c, d**), and 25–50 Ma (**e, f**). Standard model at the left, based on Coates et al. (2004) (**a, c, e**). New model at the right, based on Montes et al. (2012) (**b, d, f**)

present-day Panama and Colombia by 13–10 Ma (Montes et al. 2015; Fig. 2.7). Some biological data also suggested an emergence of the Panamanian land bridge earlier than the previous estimate of ∼3 Ma. Divergence estimates for key Neotropical plants and animals (including marine taxa) suggested that several waves of dispersal to South America occurred between the interval ∼20 and 6 Ma following Bacon et al. (2015). This challenging interpretation has been recently criticized based on the available geological and biological data (e.g., Marko et al. 2015; O'Dea et al. 2016) while acknowledging a more classical, younger age for the establishment of the Panama bridge and associated massive migration events. However, the interesting review of O'Dea et al. (2016) introduced some inaccurate data and mistakes (e.g., Fig. S3): (1) the dates for most immigration events should be taken with caution since the majority of first records of North American taxa in South America are not precisely dated (see Chaps. 4 and 5); (2), several immigration events to South America are incorrect (e.g., *Lontra* is present since the Ensenadan but not since the Lujanian; *Galictis* since the Vorohuean but not since the Ensenadan, *Canis* is only known since the latest Pleistocene but not since the Middle Pleistocene; see Chap. 4); and (3), the first record of Neotropical carnivorans has to be analyzed in the context of the bias in the SA Pleistocene fossil record (see Chap. 5 , Prevosti and Soibelzon 2012).

The picture of land bridge development is still far from conclusively resolved, and there are other possible interpretations (Coates et al. 2004; Coates and Stallard 2013, 2015; Leigh et al. 2014). Coates and Stallard (2013, 2015) argued that narrow but deep marine barriers existed between the island arc and the Americas 15–12 Ma and that the final connection was not established until ∼3 Ma. Furthermore, it is uncertain to what degree the Central American peninsula was actually subaerial at this time (Leigh et al. 2014). There is also evidence for a persistent shallow-water connection between the Pacific and the Caribbean through the 10–3.5 Ma interval, with complete interruption occurring only around 3 Ma (Coates et al. 2004; Osborne et al. 2014). Even if the interpretation of Montes et al. (2015) concerning the early connection of the Panama arc with South America is accepted, their data cannot confirm the existence of a complete terrestrial corridor between Central and South America. Equally, the divergence dates and immigration rates reported by Bacon et al. (2015) do not constitute conclusive support for an early southward irruption of North American mammals interpreted as a kind of "advance" GABI (Woodburne 2010; Cione et al. 2015). Roughly, their results are congruent with the traditional interpretation of the vertebrate fossil record, with the addition of a few dispersals in the Miocene and possibly a larger event occurring during the early and late Pliocene. Obviously, any estimates of age of clade divergences are contingent on assumptions made in the construction of models or the choice of calibrations, and the relevant cladogenetic events may have occurred outside South America.

In any case, it is clear that the new datasets indicate that the rise of the Panamanian land bridge is much more complex than previously thought and that the evanescent existence of terrestrial connections between Central and South America before 3 Ma is quite plausible. Changes in sea level, climate, and vegetation must have affected the Panamanian faunal highway in a multitude of ways over time (e.g., Leigh et al. 2014; Bagley and Johnson 2014).

Lower sea levels presumably resulted in the existence of a temporary or nearly complete isthmus in the late Miocene, with a permanently complete terrestrial bridge in place by ~3 Ma (produced by a sea level reduction related to increased northern glaciation) and later Pleistocene glaciations enhancing its permeability (Woodburne 2010; Leigh et al. 2014; Bagley and Johnson 2014). This situation was reversed during the interglacials, when higher sea levels could reconnect the Atlantic and Pacific Oceans (Bagley and Johnson 2014). Pacific dry forests in lower Central America were established in the early Pliocene (>4 Ma); remnants were present until the Late Pleistocene when these areas became savannas with patchy forest (Bagley and Johnson 2014). Northern temperate tree species were present in lowland Central America in the late Miocene–Pliocene, suggesting that the composition of tropical rainforests was recurrently affected by cooler and drier climates (Woodburne 2010). During the last glacial maximum, montane forests were found at lower altitudes. A high-altitude Paramo corridor was thus established, which could have assisted the migration of mammals adapted to grasslands and other open landscapes (Webb 1985, 1991, 2006; Woodburne 2010; Bagley and Johnson 2014; but see Colinvaux et al. (1996) for an alternative interpretation).

The classic interpretation places the GABI as having occurred subsequent to ~3.0 Ma, with pulses at 2.6–2.4 Ma (Marplatan), a major event of faunal exchange at ~1.8 Ma (early Ensenadan), followed by still other, younger events (Woodburne 2010; Prevosti and Soibelzon 2012). Some early migration events toward North America recorded around 9 Ma and toward South America at ~7.3 Ma and ~5 Ma are exemplified by ground sloths, procyonids, and cricetids, respectively (Woodburne 2010; Prevosti and Soibelzon 2012; Cione et al. 2015). Recently, several North American mammal taxa have been reported from ~9.5 Ma deposits in the Amazon basin. The presence of a dromomerycine artiodactyl, gomphotheres, peccaries, and tapirs in these contexts constitutes evidence of an earlier terrestrial interchange between North and South America (Campbell et al. 2009, 2010; Frailey and Campbell 2012; Prothero et al. 2014). However, the taxonomy of some of these fossils and, in particular, their age has been questioned (Alberdi et al. 2004; Ferretti 2008; Lucas and Alvarado 2010; Lucas 2013; Mothé and Avilla 2015). Compounding the problem is that their provenance is indeterminate in some instances (e.g., the dromomerycine artiodactyl was found in reworked deposits; see Prothero et al. 2014). Other than the aforementioned, North American taxa recorded in the South American Neogene begin with procyonids at ~8 Ma, cricetid rodents in the early Pliocene (~5.3–4 Ma), peccaries in the mid-Pliocene (~3.3 Ma), and camelids, canids, equids, gomphotheres, and weasels in the late Pliocene (2.8–2.6 Ma). The Quaternary fossil record shows a massive migration in the early Pleistocene (~1.8 Ma) with the first records of cervids, tapirs, ursids, felids, large canids, and otters (Pardiñas 1999; Prevosti et al. 2006; Woodburne et al. 2006; Prevosti and Pardiñas 2009; Woodburne 2010; Prevosti and Soibelzon 2012; Cione et al. 2015) (Fig. 2.8). Further migrations followed in the Late Pleistocene and the

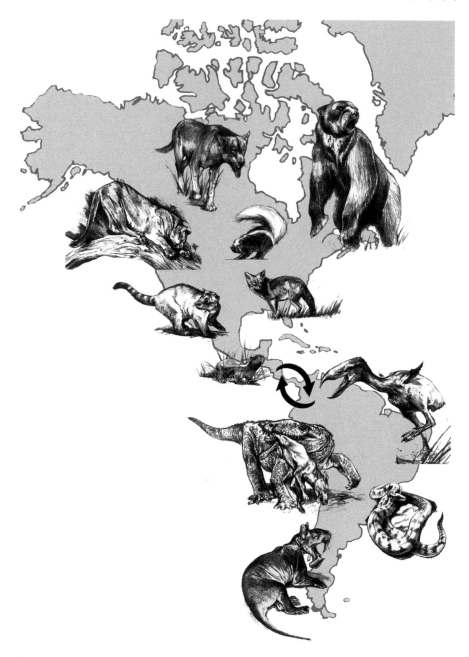

Fig. 2.8 South American carnivores. During most of the Cenozoic principal predators from the terrestrial ecosystems were marsupial relatives (Sparassodonta), terror birds (Phorusrhacidae), giant crocodiles (Sebecidae), and giant snakes (Madtsoiidae). In relation to the GABI and since the late Miocene, several placental groups (Procyonidae, Mustelidae, Mephitidae, Canidae, Ursidae, Felidae) started occupying different niches, while native predators declined. From the SA native carnivores only terror birds reached NA. Reconstructions from Jorge Blanco; carnivorans adapted from Forasiepi et al. (2007)

Holocene (Woodburne 2010), including a very important increase in carnivoran diversity ~ 12 Ka (Prevosti and Soibelzon 2012; Prevosti et al. 2013).

Migrations from South to North America occurring in the interval ~ 9–7 Ma (late Miocene) include ground sloths and terror birds (Fig. 2.8), followed by other groups of ground sloths and giant armadillos in the early Pliocene (~ 5 Ma), hydrochoerids (capybaras), megatherids, and porcupines in the late Pliocene (~ 2.7 Ma), and anteaters, toxodonts, and opossums in the Pleistocene (Woodburne 2010; Cione et al. 2015). These southern migrants proved to be less successful than their northern counterparts in South America, with limited speciation followed by extinction of most taxa of South American origin (Woodburne 2010; Carrillo et al. 2015). In contrast, northern lineages diversified in South America—sometimes spectacularly, as in the case of canids—and have now become a fully integrated part of the extant mammalian fauna of South America (Woodburne 2010; Prevosti and Soibelzon 2012; Leigh et al. 2014; Cione et al. 2015).

References

Alberdi MT, Prado JL, Salas R (2004) The Pleistocene gomphotheres (Gomphotheriidae, Proboscidea) from Peru. Neues Jahrb Geol P-A 231:423–452

Antoine P-O, Marivaux L, Croft D et al (2012) Middle Eocene rodents from Peruvian Amazonia reveal the pattern and timing of caviomorph origins and biogeography. Proc R Soc B Biol Sci 279:1319–1326

Bacon CD, Silvestro D, Jaramillo C et al (2015) Biological evidence supports an early and complex emergence of the Isthmus of Panama. Proc Natl Acad Sci 112:6110–6115

Bacon CD, Molnar P, Antonelli A, Crawford A, Montes C (2016) Quaternary glaciation and the Great American Biotic Interchange. Geology 44:375–378

Bagley JC, Johnson JB (2014) Phylogeography and biogeography of the lower Central American Neotropics: diversification between two continents and between two seas. Biol Rev 89:767–790

Barreda V, Guler V, Palazzesi L (2008) Late Miocene continental and marine palynological assemblages from Patagonia. Dev Quat Sci 11:343–350

Benedetto JL (2010) El continente de Gondwana a través del tiempo. Una introducción a la Geología Histórica. Academia Nacional de Ciencias, Córdoba

Bershaw J, Garzione CN, Higgins P et al (2010) Spatial-temporal changes in Andean plateau climate and elevation from stable isotopes of mammal teeth. Earth Planet Sci Lett 289:530–538

Bond M, Cerdeño E, López G (1995) Los ungulados nativos de América del Sur. In: Leone G, Tonni EP (eds) Alberdi MT. Monografías del Museo Nacional de Ciencias Naturales, Madrid, pp 259–275

Bond M, Tejedor MF, Campbell KE et al (2015) Eocene primates of South America and the African origins of New World monkeys. Nature 520:538–541

Borrero C, Pardo A, Jaramillo CM et al (2012) Tectonostratigraphy of the Cenozoic Tumaco forearc basin (Colombian Pacific) and its relationship with the northern Andes orogenic build up. J S Am Earth Sci 39:75–92

Brea M, Zucol AF, Iglesias A (2012) Fossil plant studies from late early Miocene of the Santa Cruz Formation: paleoecology and paleoclimatologyat the passive margin of Patagonia, Argentina. In: Vizcaíno SF, Kay RF, Bargo MS (eds) Early Miocene Paleobiology in Patagonia. Cambridge University Press, Cambridge, pp 104–128

Brea M, Zucol A, Franco MJ (2013) Paleoflora de la Formación Paraná (Mioceno Tardío), Cuenca Chaco-Paranaense. Publicación Especial APA 14:28–40

Campbell KE, Frailey CD, Romero-Pittman L (2006) The Pan-Amazonian Ucayali Peneplain, late Neogene sedimentation in Amazonia, and the birth of the modern Amazon River system. Palaeogeogr Palaeoclimatol Palaeoecol 239:166–219

Campbell KE Jr, Frailey CD, Romero-Pittman L (2009) In defense of *Amahuacatherium* (Proboscidea: Gomphotheriidae). Neues Jahrb Geol P-A 252:113–128

Campbell KE Jr, Prothero DR, Romero-Pittman L, Hertel F, Rivera N (2010) Amazonian magnetostratigraphy: dating the first pulse of the Great American Faunal Interchange. J S Am Earth Sci 29:619–626

Carrillo JD, Forasiepi AM, Jaramillo C, Sánchez-Villagra MR (2015) Neotropical mammal diversity and the Great American Biotic Interchange: a palaeontological perspective from northern South America. Front Genet—Evolutionary and Population Genetics 5(Article 451):1–11

Cione AL, Tonni EP (2001) Correlation of Pliocene to Holocene southern South American and European vertebrate-bearing units. Boll Soc Paleontol Ital 40:1–7

Cione AL, Tonni EP, Soibelzon L (2003) The broken zig-zag: late Cenozoic large mammal and tortoise extinctions in South America. Rev Mus Arg Sci Nat Bernardino Rivadavia 5:1–19

Cione AL, Tonni EP, Soibelzon L (2008) Did humans cause the Late Pleistocene-early Holocene mammalian extinctions in South America in a context of shrinking open areas? In: Haynes G (ed) American Megafaunal Extinctions at the End of the Pleistocene. Springer Science, Dordrecht, pp 125–144

Cione AL, Gasparini G, Soibelzon E et al (2015) The Great American Biotic Interchange. A South American Perspective, Springer, Dordrecht

Clapperton C (1993) Quaternary geology and geomorphology of South America. Elsevier, Amsterdam, p 779

Clark PU, Dyke AS, Shakun JD, Carlson AE, Clark J, Wohlfarth B, Mitrovica JX, Hostetler SW, McCabe AM (2009) The Last Glacial Maximum. Science 325:710–714

Coates AG, Stallard RF (2013) How old is the Isthmus of Panama? Bull Mar Sci 89:801–813

Coates A, Stallard R (2015) Historia Natural del Istmo de Panamá. In: Rodríguez Mejía F, O'Dea A (eds) Historia Natural del Istmo de Panamá. Toppan Lee Fung, Panamá, pp 17–27

Coates AG, Collins LS, Aubury MP, Berggren WA (2004) The geology of the Darien, Panama, and the late Miocene-Pliocene collision of the Panama arc with northwestern South America. Bull Geol Soc Am 116:1327–1344

Colinvaux PA, De Oliveira PE, Moreno JE, Miller MC, Bush MB (1996) A long pollen record from lowland Amazonia: forest and cooling in glacial times. Science 274:85–88

Cozzuol MA (2006) The Acre vertebrate fauna: age, diversity, and geography. J S Am Earth Sci 21:185–203

De la Fuente MS, Sterli J, Maniel IJ (2014) Origin, evolution and biogeographic history of South American turtles. Springer, Dordrecht

Del Río CJ, Griffin M, McArthur JM et al (2013) Evidence for early Pliocene and late Miocene transgressions in southern Patagonia (Argentina): 87Sr/86Sr ages of the pectinid *"Chlamys" actinodes* (Sowerby). J S Am Earth Sci 47:220–229

Dozo MT, Bouza P, Monti A et al (2010) late Miocene continental biota in Northeastern Patagonia (Península Valdés, Chubut, Argentina). Palaeogeogr Palaeoclimatol Palaeoecol 297:100–109

Erikson JP, Kelley SA, Osmolovsky P, Verosub KL (2012) Linked basin sedimentation and orogenic uplift: the Neogene Barinas basin sediments derived from the Venezuelan Andes. J S Am Earth Sci 39:138–156

Farris DW, Jaramillo C, Bayona G et al (2011) Fracturing of the Panamanian Isthmus during initial collision with: South America. Geology 39:1007–1010

Ferretti MP (2008) A review of South American gomphotheres. N Mex Nat Hist Sci Mus Bull 44:381–391

Forasiepi AM, Martinelli AG, Blanco JL (2007) Bestiario fósil. Mamíferos del pleistoceno de la Argentina. Albatros Editorial, Buenos Aires, p 192

Frailey CD, Campbell KE (2012) Two new genera of peccaries (Mammalia, Artiodactyla, Tayassidae) from upper Miocene deposits of the Amazon Basin. J Paleontol 86:852–877

Garzione CN, Hoke GD, Libarkin JC et al (2008) Rise of the Andes. Science 320:1304–1307

Goin FJ, Abello MA, Chornogubsky L (2010) Middle Tertiary marsupials from central Patagonia (early Oligocene of Gran Barranca): understanding South America's Grande Coupure. In: Madden RH, Carlini AA, Vucetich MG, Kay RF (eds) The Paleontology of Gran Barranca: evolution and environmental change through the middle Cenozoic of Patagonia. Cambridge University Press, New York, pp 71–107

Goin FJ, Woodburne MO, Zimicz AN, Martin GM, Chornogubsky L (2016) A brief history of South American metatherians, evolutionary contexts and intercontinental dispersals. Springer, Dordrecht

Gross M, Piller WE, Ramos MI, da Silva Paz Douglas, Jackson J (2011) late Miocene sedimentary environments in south-western Amazonia (Solimões Formation; Brazil). J S Am Earth Sci 32:169–181

Guerstein G, Guler M, Brinkhuis H, Wrnaar J (2010) Mid-Cenozoic paleoclimatic and paleoceanographic trends in the southwestern Atlantic Basins: a dinoflagellate view. In: Madden RH, Carlini AA, Vucetich MG, Kay RF (eds) The Paleontology of Gran Barranca: evolution and environmental change through the middle Cenozoic of Patagonia. Cambridge University Press, New York, pp 398–409

Gutiérrez-García TA, Vázquez-Domínguez E (2013) Consensus between genes and stones in the biogeographic and evolutionary history of Central America. Quat Res 79:311–324

Haq BU, Hardenbol J, Vail PR et al (1987) Chronology of fluctuating sea levels since the Triassic. Science 235:1156–1167

Haug GH, Tiedemann R (1998) Effect of the formation of the Isthmus of Panama on Atlantic Ocean thermohaline circulation. Nature 393:673–676

Helmens K, Van der Hammen T (1994) The Pliocene and Quaternary of the high plain of Bogotá (Colombia): a history of tectonic uplift, basin development and climatic change. Quat Int 22:41–81

Hoorn C, Wesselingh FP, ter Steege H et al (2010) Amazonia through time: Andean uplift, climate change, landscape evolution, and biodiversity. Science 330:927–931

Hovikoski J, Räsänen M, Gingras M et al (2007) Palaeogeographical implications of the Miocene Quendeque Formation (Bolivia) and tidally-influenced strata in southwestern Amazonia. Palaeogeogr Palaeoclimatol Palaeoecol 243:23–41

Iturralde-Vinent MA, MacPhee RDE (1999) Paleogeography of the Caribbean region: implications for Cenozoic biogeography. Bull Am Mus Nat Hist 238:1–95

Kraglievich JL (1952) El perfil geológico de Chapadmalal y Miramar. Resumen Preliminar. Rev Mus Mun Cs Nat y Trad Mar del Plata 1:8–37

Latrubesse EM, Cozzuol M, da Silva-Caminha SAF et al (2010) The late Miocene paleogeography of the Amazon Basin and the evolution of the Amazon River system. Earth-Science Rev 99:99–124

Leanza HA, Hugo CA (1997) Hoja geológica 3969-III Picun Leufú. Bol SEGEMAR 218:1–135

Leigh EG, O'Dea A, Vermeij GJ (2014) Historical biogeography of the Isthmus of Panama. Biol Rev 89:148–172

Livermore R, Hillenbrand C-D, Meredith M, Eagles G (2007) Drake Passage and Cenozoic climate: an open and shut case? Geochem Geophys 8:1–11

Lucas SG (2013) The palaeobiogeography of South American gomphotheres. J Palaeogeogr 2:19–40

Lucas SG, Alvarado GE (2010) Fossil Proboscidea of the upper Cenozoic of Central America: taxonomy, evolutionary and paleobiogeographic significance. Rev Geol Am Cent 42:9–42

MacFadden BJ, Cerling TE, Prado J (1996) Cenozoic terrestrial ecosystem evolution in Argentina: evidence from carbon isotopes of fossil mammal teeth. Palaios 11:319–327

Marengo H (2015) Neogene micropaleontology and stratigraphy of Argentina. The Chaco-Paranense basin and the Península de Valdés, Springer, Dordrecht

Marko PB, Eytan RI, Knowlton N (2015) Do large molecular sequence divergences imply an early closure of the Isthmus of Panama? Proc Natl Acad Sci U S A 112:E5766. doi:10.1073/pnas. 1515048112

Marshall LG, Cifelli RL (1990) Analysis of changing diversity patterns in Cenozoic Land Mammal Age faunas, South America. Palaeovertebrata 19:169–210

Meng J, McKenna MC (1998) Faunal turnovers of Palaeogene mammals from the Mongolian Plateau. Nature 394:364–367

Montes C, Cardona A, McFadden R et al (2012) Evidence for middle Eocene and younger land emergence in central Panama: implications for Isthmus closure. Bull Geol Soc Am 124:780–799

Montes C, Cardona A, Jaramillo C et al (2015) middle Miocene closure of the Central American Seaway. Science 348:226–230

Mothé D, Avilla L (2015) Mythbusting evolutionary issues on South American Gomphotheriidae (Mammalia: Proboscidea). Quat Sci Rev 110:23–35

O'Dea A, Lessios H, Coates A, Eytan R, Restrepo-Moreno S, Cione A, Collins L, de Queiroz A, Farris D, Norris R, Stallard R, Woodburne M, Aguilera O, Aubry M, Berggren W, Budd A, Cozzuol M, Coppard S, Duque-Caro S, Finnegan S, Gasparini G, Grossman E, Johnson K, Keigwin L, Knowlton N, Leigh E, Leonard-Pingel J, Marko P, Pyenson N, Rachello-Dolmen P, Soibelzon E, Soibelzon L, Todd J, Vermeij G, Jackson J (2016) Formation of the Isthmus of Panama. Sci Adv Sci Adv 2:e1600883

Ochoa D, Hoorn C, Jaramillo C et al (2012) The final phase of tropical lowland conditions in the axial zone of the Eastern Cordillera of Colombia: evidence from three palynological records. J S Am Earth Sci 39:157–169

Ortiz Jaureguizar E, Prado JL, Alberdi MT (1995) Análisis de las comunidades de mamíferos continentales del Plio-Pleistoceno de la región pampeana y su comparación con la del área demediterráneo occidental. In: Alberdi MT, Leone G, Tonni EP (eds) Evolución Biológica y Climática de la Región Pampeana durante los Últimos Cinco Millones de Años. Un Ensayo de Correlación con el Mediterráneo Occidental. Monografías delMuseo Nacional de Ciencias Naturales, Madrid, pp 385–406

Osborne AH, Newkirk DR, Groeneveld J, Martin EE, Tiedemann R, Frank M (2014) The seawater neodymium and lead isotope record of the final stages of Central American Seaway closure. Paleoceanography 29:715–729

Palazzesi L, Barreda VD, Cuitiño JI et al (2014) Fossil pollen records indicate that Patagonian desertification was not solely a consequence of Andean uplift. Nat Commun 5:3558

Pardiñas UFJ (1999) Fossil murids: taxonomy, paleoecology, and paleoenvironments. Quat South Am 12:225–254

Pascual R, Ortiz Jaureguizar E (1990) Evolving climates and mammal faunas in Cenozoic South American. J Human Evol 19:23–60

Patterson BD, Solari S, Velazco PM (2012) The role of the Andes in the diversification and biogeography of Neotropical mammals. In: Patterson BD, Costa LP (eds) Bones, clones, and biomes: the history and geography of Recent Neotropical mammals. University of Chicago Press, Chicago, pp 351–378

Pérez LM (2013) Nuevo aporte al conocimiento de la edad de la Formación Paraná, Mioceno de la Provincia de Entre Ríos, Argentina. Publicación Especial APA 14:7–12

Pfuhl HA, McCave IN (2005) Evidence for late Oligocene establishment of the Antarctic Circumpolar Current. Earth Planet Sci Lett 235:715–728

Ponce JF, Rabassa J, Coronato A, Borromei AM (2011) Palaeogeographical evolution of the Atlantic coast of Pampa and Patagonia from the last glacial maximum to the middle Holocene. Biol J Linn Soc 103:363–379

Prevosti FJ, Pardiñas UFJ (2009) Comment on "The oldest South American Cricetidae (Rodentia) and Mustelidae (Carnivora): late Miocene faunal turnover in central Argentina and the Great American Biotic Interchange" by D.H. Verzi and C.I. Montalvo. Palaeogeogr Palaeoclimatol Palaeoecol 280:543–547

Prevosti F, Soibelzon L (2012) Evolution of the South American carnivores (Mammalia, Carnivora): a paleontological perspective. In: Patterson BD, Costa LP (eds) Bones, clones, and biomes: the history and geography of Recent Neotropical mammals. University of Chicago Press, Chicago, pp 102–122

Prevosti FJ, Gasparini GM, Bond M (2006) On the systematic position of a specimen previously assigned to Carnivore from the Pliocene of Argentina and its implications for the Great American Biotic Interchange. Neues Jahrb Geol P-A 242:133–144

Prevosti FJ, Forasiepi A, Zimicz N (2013) The evolution of the Cenozoic terrestrial mammalian predator guild in South America: competition or replacement? J Mamm Evol 20:3–21

Prothero DR (1994) The Eocene-Oligocene transition: paradise lost. Columbia Univ. Press, New York

Prothero DR, Campbell KE, Beatty BL, Frailey CD (2014) New late Miocene dromomerycine artiodactyl from the Amazon Basin: implications for interchange dynamics. J Paleontol 88:434–443

Rabassa J (2008) Late Cenozoic glaciations in Patagonia and Tierra del Fuego. Dev Quat Sci 11:151–204

Rabassa J, Ponce JF (2013) The Heinrich and Dansgaard-Oeschger climatic events during Marine Isotopic Stage 3: searching for appropriate times for human colonization of the Americas. Quat Int 299:94–105

Ramos VA (1999) Plate tectonic setting of the Andean Cordillera. Episodes 22:183–190

Ramos VA (2009) Anatomy and global context of the Andes: main geologic features and the Andean orogenic cycle. Geol Soc Am Mem 204:31–65

Räsänen ME, Linna AM, Santos JCR, Negri FR (1995) Late Miocene tidal deposits in the Amazonian foreland basin. Science 269:386–390

Reguero M, Goin F, Acosta C, Tania H (2013) Late Cretaceous/Paleogene West Antarctica terrestrial biota and its intercontinental affinities. Springer, Dordrecht

Rutter N, Coronato A, Helmens K et al (2012) Glaciations in North and South America from the Miocene to the Last Glacial Maximum. Comparisons, linkages and uncertainties. Springer, Dordrecht

Scher HD, Martin EE (2006) The timing and climatic influence of the opening of Drake passage. Science 312:428–430

Schultz P, Zarate M, Hames W et al (1998) A 3.3 Ma impact in Argentina and possible consequences. Science 282:2061–2063

Simpson GG (1950) History of the fauna of Latin America. Am Sci 38:361–389

Simpson GG (1980) Splendid Isolation. The curious history of South American mammals. Yale University Press, New Haven

Sluijs A, Brinkhuis H, Schouten S, Bohaty SM, John CM, Zachos JC, Sinninghe Damste JS, Crouch EM, Dickens GR (2007) Environmental precursors tolight carbon input at the Paleocene/Eocene boundary. Nature 450:1218–1221

Starck D, Anzótegui LM (2001) The late climatic change persistence of a climatic signal through the orogenic stratigraphic record in northwestern of Argentina. J S Am Earth Sci 14:763–774

Stehli FG, Webb SD (1985) The Great American. Biotic Interchange. Plenum Press, New York

Tambussi C, Degrange F (2013) South American and Antarctic continental Cenozoic birds. Paleobiogeographic affinities and disparities. Springer, Dordrecht

Tonni EP, Alberdi MT, Prado JL, Bargo MS, Cione AL (1992) Changes of mammal assemblages in the Pampean Region (Argentina) and their relation with the Plio-Pleistocene boundary. Palaeogeogr Palaeoclimatol Palaeoecol 95:179–194

Uba CE, Hasler CA, Buatois LA et al (2009) Isotopic, paleontologic, and ichnologic evidence for late Miocene pulses of marine incursions in the central Andes. Geology 37:827–830

Van der Hammen T (1973) Upper Quaternary vegetational and climatic sequence of the Fúquene area (Eastern, cordillera, Colombia). Palaeogeogr Palaeoclimatol Palaeoecol 14:9–92

Vizcaíno SF, Fariña RA, Zárate M et al (2004) Palaeoecological implications of the mid-Pliocene faunal turnover in the Pampean Region (Argentina). Palaeogeogr Palaeoclimatol Palaeoecol 213:101–113

Webb SD (1985) Late Cenozoic mammal dispersals between the Americas. In: Stehli FG, Webb SD (eds) The Great American Biotic Interchange. Plenum Press, New York, pp 357–386

Webb SD (1991) Ecogeography and the Great American Interchange. Paleobiology 17:266–280

Webb SD (2006) The Great American Biotic Interchange: Patterns and Processes. Ann Missouri Bot Gard 93:245–257

Wijninga VM (1996) Palynology and paleobotany of the early Pliocene section Río Frío 17 (Cordillera Oriental, Colombia): Biostratigraphical and chronostratigraphical implications. Rev Palaeobot Palynol 92:329–350

Wilf P, Cúneo R, Escapa IH, Pol D, Woodburne MO (2013) Splendid and seldom isolated: the paleobiogeography of Patagonia. Annu Rev Earth Planet Sci 41:561–603

Wing SL, Alroy J, Hickey LJ (1995) Plant and mammal diversity in the Paleocene to early Eocene of the Bighorn Basin. Palaeogeogr Palaeoclimatol Palaeoecol 115:117–156

Woodburne MO (2010) The Great American Biotic Interchange: dispersals, tectonics, climate, sea level and holding pens. J Mamm Evol 17:245–264

Woodburne MO, Cione AL, Tonni EP (2006) Central American provincialism and the Great American Biotic Interchange. In: Carranza-Castañeda Ó, Lindsay EH (eds) Advances in late Tertiary vertebrate paleontology in Mexico and the Great American Biotic Interchange. Universidad Nacional Autónoma de México, Instituto de Geología and Centro de Geociencias, Publicación Especial 4, México, pp 73–101

Woodburne MO, Goin FJ, Bond M et al (2014a) Paleogene land mammal faunas of South America; a response to global climatic changes and indigenous floral diversity. J Mamm Evol 21:1–73

Woodburne MO, Goin FJ, Raigemborn MS, Heizler M, Gelfo JN, Oliveira EV (2014b) Revised timing of the South American early Paleogene land mammal ages. J S Am Earth Sci 54:109–119

Yrigoyen MR (1979) Cordillera Principal. Actas II Simp Geol Reg, Acad Nac Cs Córdoba 1:651–694

Zachos J, Pagani M, Sloan L et al (2001) Trends, global rhythms, aberrations in global climate 65 Ma to Present. Science 292:686–693

Chapter 3
South American Endemic Mammalian Predators (Order Sparassodonta)

Abstract The Sparassodonta was a clade of mammalian predators that evolved in South America from the early Paleocene (?Tiupampan–Peligran) or early Eocene (Itaboraian) to the early Pliocene (Chapadmalalan). They were a monophyletic group of metatherians closely related to living marsupials (e.g., opossums and kangaroos). Diverse ecological niches presented many opportunities for occupation by different morphotypes (principally defined by body mass and locomotion). The probable diet was hypercarnivorous for about 90% of the nearly 60 currently identified species. Here, we present a synthesis of the systematics, distribution, and paleoecology of the extinct Sparassodonta.

Keywords Borhyaenidae · Hathliacynidae · Hondadelphydae · Proborhyaenidae Thylacosmilidae

3.1 Introduction

The Sparassodonta was a group of predaceous metatherians, now extinct. Their fossil record arguably extends from the early Paleocene (possibly ?Tiupampan–Peligran) or early Eocene (Itaboraian) up to the early Pliocene (Chapadmalalan) (Simpson 1950, 1980; Marshall 1977a, 1978, 1979, 1981; Goin and Pascual 1987; Marshall and de Muizon 1988; de Muizon 1994, 1998, 1999; Forasiepi 2009; Babot and Forasiepi 2016) (Table 3.1). Sparassodonta is an exclusively South American monophyletic group with most fossil evidence concentrated in the southern parts of the continent (Fig. 3.1).

The phylogenetic affinities of Sparassodonta within Metatheria have been a controversial issue in recent decades with a number of opposing positions proposed. Earlier views that Sparassodonta were directly related to one or more of thylacinids, dasyurids, and didelphids have been challenged by recent work showing that sparassodonts cannot be included in crown group Marsupialia (e.g., Szalay 1994; Rougier et al. 1998, 2004; Babot 2005; Ladevèze and de Muizon 2007, 2010; Forasiepi 2009; Engelman and Croft 2014; Forasiepi et al. 2015; Suarez et al. 2015;

© Springer International Publishing AG 2018
F.J. Prevosti and A.M. Forasiepi, *Evolution of South American Mammalian Predators During the Cenozoic: Paleobiogeographic and Paleoenvironmental Contingencies*, Springer Geology, https://doi.org/10.1007/978-3-319-03701-1_3

Table 3.1 Distribution of sparassodonts in the South American Stages/Ages

Taxa	Tiupampan	Peligran	Itaboraian	Riochican	Casamayoran	Mustersan	Tinguirirican	Deseadan	Colhuehuapian	Santacrucian	Friasian	Colloncuran	Laventan	Mayoan	Chasicoan	Huayquerian	Montehermosan	Chapadmalalan
Nemolestes spalacotherinus					X													
Patene campbelli						X												
Patene coluapiensis					X													
Patene simpsoni			X															
Stylocynus paranensis																X		
Hondadelphys fieldsi													X					
Acyon myctoderos													X					
Acyon tricuspidatus										X								
Acyon herrerae (nomen dubium)								X										
Borhyaenidium altiplanicus																	X	
Borhyaenidium musteloides															X			
Borhyaenidium riggsi																		X
Chasicostylus castroi															X			
Cladosictis centralis									X									
Cladosictis patagonica										X								
Contrerascynus borhyaenoides																	X	
Notictis ortizi													X					
Notocynus hermosicus																	X	
Notogale mitis								X										
Perathereutes pungens										X								
Pseudonotictis chubutensis											X							
Pseudonotictis pusillus										X								
Sallacyon hoffstetteri								X										
Sipalocyon externa									X									
Sipalocyon gracilis										X								
Sipalocyon obusta (nomen dubium)										X								
Procladosictis anomala						X												
Dukecynus magnus													X					
Lycopsis longirostrus													X					
Lycopsis padillai											X							
Lycopsis torresi															X			
Lycopsis viverensis															X			
Pharsophorus lacerans								X										
Pharsophorus tenax								X										
Plesiofelis schlosseri								X										
Prothylacynus patagonicus										X								
Pseudolycopsis cabrerai																X		
Pseudothylacynus rectus									X									
Arminiheringia auceta					X													
Arminiheringia contigua					X													
Arminiheringia cultrata					X													
Callistoe vincei					X													
Paraborhyaena boliviana								X										
Proborhyaena gigantea								X										
Anachlysictis gracilis													X					
Patagosmilus goini												X						
Thylacosmilus atrox																X	X	X
Acrocyon riggsi										X								
Acrocyon sectorius										X								
Arctodictis munizi								X										
Arctodictis sinclairi									X									
Australohyaena antiquua								X										
Borhyaena macrodonta									X									
Borhyaena tuberata										X								
Fredszalaya hunteri								X										
Angelocabrerus daptes (nomen dubium)					X													
Argyrolestes peralestinus (nomen dubium)					X													
Eutemnodus americanus (nomen dubium)																X		
? Sparassodonta																		
Allqokirus australis	X																	
Andinodelphys cohcabanbensis	X																	
Jaskhadelphys minutus	X																	
Mayulestes ferox	X																	
Pucadelphys andinus	X																	

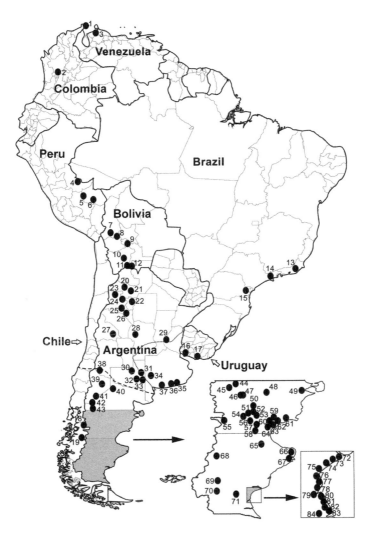

Babot and Forasiepi 2016; Fig. 3.2). The discovery of the fine skeleton of *Mayulestes ferox* from the Paleocene of Bolivia (claimed to be the most basal sparassodont; de Muizon 1994, 1998; de Muizon et al. 2015) reignited interest in the phylogeny of the group. The phylogenetic position of *Mayulestes* within Sparassodonta, as well as other Tiupampan metatherians, is matter of discussion (see below).

The nearly 60 currently identified members of the clade cover many different shapes, sizes, and locomotor types (Table 3.2). All forms fit within carnivory, most of them consistent with hypercarnivory (Marshall 1977a, 1978, 1979, 1981; Prevosti et al. 2013; Wroe et al. 2013), and with bite forces higher than placental carnivorans (Wroe et al. 2004b; Blanco et al. 2011). Locomotor habits range from terrestrial to arboreal. As indicated by Argot and Babot (2011), terrestriality appears

◄**Fig. 3.1** Map of South America with the localities where sparassodont remains were found. COLOMBIA: 1, La Guajira; 2, La Venta. VENEZUELA: 3, Urumaco. PERU: 4, Santa Rosa; 5, Fitzcarrald; 6, Madre de Dios. BOLIVIA: 7, Achiri; 8, Salla-Lurivay; 9, Tiupampa; 10, Cerdas; 11, Nazareno; 12, Quebrada Honda. BRAZIL: 13, São Jasé de Itaboraí; 14, Tremembé; 15, Curitiba. URUGUAY: 16, Arazatí; 17, Paso del Cuello. CHILE: 18, Alto Río Cisnes; 19, Pampa Castillo. ARGENTINA: 20, Estrecho del Tronco; 21, Pampa Grande; 22, Tafí Viejo; 23, Antofagasta de la Sierra; 24, Chiquimil; 25, Puerta del Corral Quemado; 26, Andalgalá; 27, Loma de las Tapias; 28, Nono; 29, Paraná; 30, Telén; 31, El Guanaco; 32, Quehué; 33, Salinas Grandes de Hidalgo; 34, Arroyo Chasicó; 35, Barranca de los Lobos; 36, Chapadmalal; 37, Monte Hermoso; 38, Quebrada Fiera; 39, Cerro Bandera and Sierra del Portezuelo Norte; 40, Paso Córdova; 41, Cañadón del Tordillo; 42, Pilcaniyeu; 43, Río Chico; 44, El Petiso; 45, Cerro Zeballos; 46, Laguna Fría; 47, La Barda; 48, Sacanana; 49, Gaiman; 50, Laguna de la Bombilla; 51, Laguna Payahilé; 52, Gran Hondonada; 53, Rinconada de López; 54, La Curandera; 55, Río Senguer; 56, Cerro del Humo; 57, Barranca Norte (north slope of the Colhué-Huapi Lake); 58, Barranca Sur or Gran Barranca (south slope of the Colhué-Huapi Lake); 59, Cabeza Blanca; 60, Yacimiento Las Flores; 61, Cerro Redondo; 62, Punta Peligro; 63, Bajo de la Palangana; 64, Cañadón Hondo; 65, Pico Truncado; 66, Puerto Deseado; 67, La Flecha; 68, Lago Pueyrredón; 69, Sheuen; 70, Karaiken; 71, Río Santa Cruz; 72, Monte León; 73, La Cueva; 74, Yegua Quemada; 75, Monte Observación; 76, Jack Harvey; 77, Cañadón de las Vacas; 78, Wreck Hat; 79, Coy Inlet; 80, La Costa; 81, Corrigen Kaik; 82, Estancia Angelina; 83, Río Gallegos (locality); 84, Kallik Aike Norte (=Felton's Estancia)

to have been the primitive condition for the group (recorded in *Arctodictis sinclairi*, *Borhyaena tuberata*, *Callistoe vincei*, *Lycopsis longirostrus*, *Thylacosmilus atrox*, and in the stem marsupials from Tiupampa, with a gradient of increasing arboreality from *Pucadelphys andinus*, *Andinodelphys cochabambensis*, to *Mayulestes ferox*; de Muizon and Argot 2003; Argot and Babot 2011). The climbing abilities among sparassodonts apparently evolved independently in different lineages of medium- to small-sized   hathliacynids   (*Cladosictis   patagonica*,   *Sipalocyon   gracilis*, *Pseudonotictis pusillus*) and some large-size borhyaenoids (*Prothylacynus patagonicus*). An incipient cursoriality has been suggested for *Borhyana tuberata* and *Thylacosmilus atrox* (Argot 2004a; Ercoli et al. 2012).

Both the total taxonomic diversity of Sparassodonta through the Cenozoic and their morphological disparity seem to be lower than in placental Carnivora, which fits with the general pattern of a more restricted morphospace occupied by marsupials compared to placentals (Sears 2004; Sánchez-Villagra 2013; Prevosti et al. 2012a; Echarri and Prevosti 2015; a contrary conclusion in Goswami et al. 2011). When analyzing the taxonomic composition of the most studied SA fossil association—Santa Cruz Formation, early Miocene (Santacrucian)—, it may be seen that the sparassodont diversity reached eleven (or at least nine) species, which roughly matches carnivoran taxonomic richness in some current environments (Prevosti et al. 2012b, but see Croft 2001, 2006; Wroe et al. 2004a). In addition, during most of the Cenozoic, the carnivorous adaptive zone of South America was shared between the Sparassodonta and nonmammalian taxa, such as Sebecidae (crocodiles), Phorusrhacidae ("terror birds"), Madtsoiidae (snakes), and, for about 3 Ma (from the late Miocene to the mid-Pliocene), with placental carnivores and didelphimorphian marsupials (Degrange et al. 2010; Prevosti et al. 2013; Scheyer et al.

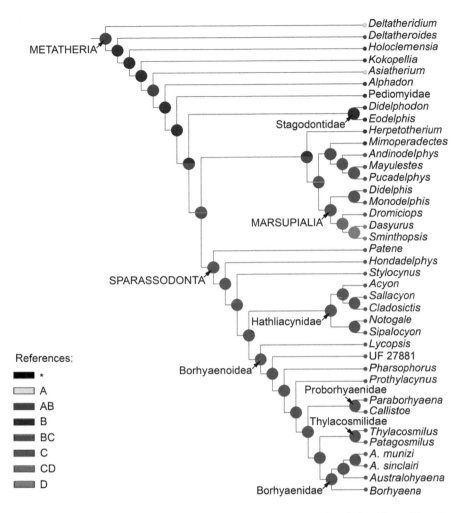

Fig. 3.2 Phylogenetic hypothesis of the Sparassodonta and their relationships with other metatherians. Cladogram obtained under implied weighting (Forasiepi et al. 2015) showing the reconstruction of the ancestral areas at each node obtained by S-DIVA (Statistical Dispersal–Vicariance Analysis) and exported from RASP (Reconstruct Ancestral State in Phylogenies). Color key represents possible ancestral ranges at different nodes: A, Asia; B, North America; C, South America; D, Australia; black with an asterisk represents other ancestral ranges

2013; Zimicz 2014). This would have favoured a partitioning of the carnivorous adaptive zone (Wroe et al. 2004a; Ercoli et al. 2013; Forasiepi and Sánchez-Villagra 2014) in which each taxonomic group would have occupied a particular role in the terrestrial ecosystems.

Table 3.2 Body mass (BM) and diet of sparassodonts

Taxa	BM (kg)	Mass category	RGA	Diet category	Comments
Nemolestes spalacotherinus	5.72	S	Unknown m4	hyp	BM from Zimicz (2012) and Prevosti et al. (2013). RGA of m3: 0.35. Diet from Prevosti et al. (2013)
Patene campbelli	1	S	Unknown m4	omn	BM from Zimicz (2012). Diet taken from *Patene simpsoni*, following Prevosti et al. (2013)
Patene coluapiensis	3.07	S	Unknown m4	omn	BM mean from Zimicz (2012) and Prevosti et al. (2013). Diet taken from *Patene simpsoni*, following Prevosti et al. (2013)
Patene simpsoni	1.35	S	Unknown m4	omn	BM mean from Zimicz (2012) and Prevosti et al. (2013). RGA of m3: 0.76. Diet from Prevosti et al. (2013)
Stylocynus paranensis	31.05	L	0.61	omn	BM from mean of tooth variables from Wroe et al. (2004a) and Prevosti et al. (2013). Diet from Prevosti et al. (2013)
Hondadelphys fieldsi	3.7	S	0.63	omn	BM from tooth variables and diet from Prevosti et al. (2013)
Acyon myctoderos	13–17.5	M to L	0.27	hyp	BM from postcranial variables from Engelman et al. 2015. Diet from Prevosti et al. (2013)
Acyon tricuspidatus	6.51	S	0.30	hyp	BM mean from Wroe et al. (2004a), Zimicz (2012), and Prevosti et al. (2013). Diet from Prevosti et al. (2013)
Acyon herrerae (nomen dubium)	8.23	M	0.30	hyp	BM mean from Wroe et al. (2004a), Zimicz (2012), and Prevosti et al. (2013). Diet from Prevosti et al. (2013)
Borhyaenidium altiplanicus	1.16	S	Unknown m4	hyp	BM from Prevosti et al. (2013). Diet taken from *Borhyaenidium musteloides*, following Prevosti et al. (2013)

(continued)

Table 3.2 (continued)

Taxa	BM (kg)	Mass category	RGA	Diet category	Comments
Borhyaenidium musteloides	1.56	S	0.30	hyp	BM and diet from Prevosti et al. (2013)
Borhyaenidium riggsi	1.98	S	0.32	hyp	BM and diet from Prevosti et al. (2013)
Chasicostylus castroi	8.27	M	Unknown m4	hyp	BM mean from Wroe et al. (2004a) and Prevosti et al. (2013). Diet taken from *B. musteloides*, following Prevosti et al. (2013)
Cladosictis centralis	3.9	S	Unknown m4	hyp	BM mean from Wroe et al. (2004a) and Prevosti et al. (2013). Diet taken from *C. patagonica*, following Prevosti et al. (2013)
Cladosictis patagonica	6.6	S	0.17	hyp	BM from Ercoli and Prevosti (2011). Diet from Prevosti et al. (2013)
Contrerascynus borhyaenoides	12.6	M	0.27	hyp	This work. BM from equations from Gordon (2003), based on the length of m3. Diet from Prevosti et al. (2013)
Notictis ortizi	1.17	S	Unknown m4	hyp	BM from Prevosti et al. (2013). Diet taken from *Pseudonotictis pusillus*, following Prevosti et al. (2013)
Notocynus hermosicus	2.48	S	Unknown m4	hyp	BM mean from Wroe et al. (2004a) and Prevosti et al. (2013). Diet taken from *S. gracilis*, following Prevosti et al. (2013)
Notogale mitis	3.06	S	0.37	hyp	BM mean from Wroe et al. (2004a), Zimicz (2012), and Prevosti et al. (2013). Diet from Prevosti et al. (2013)
Perathereutes pungens	1.75	S	0.34	hyp	BM mean from Wroe et al. (2004a) and Prevosti et al. (2013). Diet from Prevosti et al. (2013).
Pseudonotictis chubutensis	0.89	S	Unknown m4	hyp	BM from Zimicz (2014). Diet taken from *Pseudonotictis pusillus*, following Prevosti et al. (2013)

(continued)

Table 3.2 (continued)

Taxa	BM (kg)	Mass category	RGA	Diet category	Comments
Anachlysictis gracilis	17	L	0	hyp	BM mean from Wroe et al. (2004a) and Prevosti et al. (2013). Diet from Prevosti et al. (2013)
Patagosmilus goini	~16	L	Unknown m4	hyp	This work. BM from equations of upper molars from Gordon (2003). Diet from *A. gracilis,* following Prevosti et al. (2013)
Thylacosmilus atrox	117.4	L	0	hyp	BM from Ercoli and Prevosti (2011). Diet from Prevosti et al. (2013)
Acrocyon riggsi	21.65	L	0	hyp	BM mean from Wroe et al. (2004a) and Prevosti et al. (2013). Diet from Prevosti et al. (2013)
Acrocyon sectorius	22.48	L	0	hyp	BM mean from Wroe et al. (2004a) and Prevosti et al. (2013). Diet from Prevosti et al. (2013)
Arctodictis munizi	43.87	L	0	hyp	BM mean from Wroe et al. (2004a), Vizcaíno et al. (2010), and Prevosti et al. (2013). Diet from Prevosti et al. (2013)
Arctodictis sinclairi	40	L	0	hyp	BM from Ercoli and Prevosti (2011). Diet from Prevosti et al. (2013)
Australohyaena antiquua	67	L	0	hyp	BM and diet from Forasiepi et al. (2015)
Borhyaena macrodonta	32.97	L	0	hyp	BM mean from Wroe et al. (2004a) and Prevosti et al. (2013). Diet from Prevosti et al. (2013)
Borhyaena tuberata	36.4	L	0	hyp	BM from Ercoli and Prevosti (2011). Diet from Prevosti et al. (2013)
Fredszalaya hunteri	31.8	L	Unknown m4	hyp	BM and diet taken from *Prothylacynus patagonicus,* following Prevosti et al. (2013)
Angelocabrerus daptes (nom. dub.)	~20.5	L	0	hyp	BM from Zimicz (2012). Diet from Prevosti et al. (2013)

(continued)

Table 3.2 (continued)

Taxa	BM (kg)	Mass category	RGA	Diet category	Comments
Argyrolestes peralestinus (nom. dub.)	4.95	S	Not found	hyp	BM and diet taken from *Nemolestes spalacotherinus*, following Prevosti et al. (2013)
Eutemnodus americanus (nom. dub.)	36.4	L	Unknown m4	hyp	BM and diet taken from *Borhyaena tuberata*, following Prevosti et al. (2013)

RGA: relative grinding area of lower carnassial (m4); nom. dub.: nomen dubium; L: large; M: medium; S: small; hyp: hypercarnivore; omn: omnivore; meso: mesocarnivore

3.2 Systematics, Distribution, and Paleoecology

We present a synthesis of the diversity, paleoecology, and temporal and geographic distribution of the Sparassodonta Detailed paleoecological studies are limited to those species best represented in the fossil record. Where there was insufficient evidence, closely related species are also used to supply paleoecological inferences. Body mass estimates are the most accurate when estimated on the basis of postcranial variables (e.g., Millien and Bovy 2010), but this evidence is mostly unavailable, so equations based on tooth measurements provide the only possible method (Myers 2001 in Wroe et al. 2004a; Gordon 2003 in Zimicz 2012, 2014; Prevosti et al. 2013). For the hypercarnivorous sparassodonts without grinding surfaces in the molars, the body mass predictions are usually underestimates (Prevosti et al. 2012b). Values in Table 3.2 were calculated on the basis of means derived from (1) postcranial data with correction for phylogeny, (2) postcranial data without correction for phylogeny, (3) combinations of skull and tooth variables, and (4) tooth variables alone. The diet was calculated with RGA tooth equations (modified from Van Valkenburgh 1991), and values were taken from Zimicz (2012) and Prevosti et al. (2013).

Mammalia Linnaeus 1758
Metatheria Huxley 1880
Sparassodonta Ameghino 1894

Sparassodonta includes hathliacynids, borhyaenoids, and all the taxa with a closer relationship to them than to living marsupials (Forasiepi 2009).

Nemolestes Ameghino 1902a
(Tables 3.1 and 3.2)

Included species. *Nemolestes spalacotherinus* Ameghino 1902a.

Temporal and geographic distribution. Middle to late Eocene (?Itaboraian and Casamayoran), Argentina and Brazil.

Paleoecology. Small-sized sparassodont (6.5 kg or ~5 kg; Zimicz 2012; Prevosti et al. 2013, respectively). Lower molars have reduced grinding areas; RGA tooth equations suggested hypercarnivorous diet (Zimicz 2012; Prevosti et al. 2013).

Comments. *Nemolestes spalacotherinus* is known by isolated teeth from Patagonia, Argentina. In addition, one lower molar from Itaboraí, Brazil, was referred to cf *Nemolestes* sp. (Marshall 1978; Bergqvist et al. 2006). Affinities with borhyaenoids have been suggested (e.g., Marshall 1978); however, its generalized tooth morphology likely indicates that *Nemolestes* does not belong to any major division within the group (Forasiepi 2009; Fig. 3.2).

Patene Simpson 1935
(Tables 3.1 and 3.2)

Included species. *Patene coluapiensis* Simpson 1935 (type species), *P. simpsoni* Paula Couto 1952, and *P. campbelli* Goin and Candela 2004.

Temporal and geographic distribution. *Patene simpsoni* comes from the early to middle Eocene (Itaboraian and Casamayoran), Argentina and Brazil; *P. coluapiensis* from the middle Eocene (Casamayoran), Argentina; *P. campbelli* from the latest middle–earliest late Eocene (Mustersan), Peru.

Paleoecology. *Patene* includes species of small body size (*P. simpsoni*: 1.3 kg or 1.4 kg; *P. coluapiensis*: 3.6 kg or 2.54 kg; *P. campbelli*: 1 kg or 5.37 kg in Zimicz 2012 and Prevosti et al. 2013, respectively). Estimations of *P. campbelli* in Zimicz (2012) better reflect the linear measurements for the smallest of the species of the genus (Goin and Candela 2004). Their broader molar grinding areas provide RGA values on the order of that of omnivorous/mesocarnivorous taxa (Zimicz 2012; Prevosti et al. 2013).

Comments. Material assigned to *Patene* species includes isolated teeth or tooth series associated with fragmentary maxillae and dentaries (Simpson 1935, 1948; Marshall 1981; Goin et al. 1986). *Patene* was considered a hathliacynid (e.g., Marshall 1981; Goin and Candela 2004); however, cladistic analyses recovered this taxon outside the major sparassodont groups (Forasiepi 2009; Engelman and Croft 2014; Forasiepi et al. 2015; Fig. 3.2).

Patene simpsoni and *P. campbelli* have been found in tropical latitudes of South America: the first from São José de Itaboraí (Brazil) and Estrecho del Tronco, Salta (Argentina), the latter from Santa Rosa (Peru) (Goin et al. 1986; Goin and Candela 2004), whereas *P. coluapiensis* comes from higher Patagonian latitudes (Marshall 1981).

The age of the Paleogene Santa Rosa fossil site in Peru, where *P. campbelli* comes from, has been controversial, with recent views in agreement that it is late middle Eocene or late Eocene (Bond et al. 2015).

Stylocynus Mercerat 1917
(Fig. 3.3a; Tables 3.1 and 3.2)

(a)   **(b)**

Fig. 3.3 *Stylocynus paranensis* (MLP 11-94, Holotype) from the late Miocene (Huayquerian Age), Paraná (Argentina), incomplete left dentary in lateral view (**a**); *Borhyaenidium musteloides* (MLP 57-X-10-153, Holotype) from the late Miocene (Huayquerian Age), Salinas Grandes de Hidalgo (Argentina), left dentary in lateral view (**b**). Scale bar = 3 cm

Included species. *Stylocynus paranensis* Mercerat 1917.

Temporal and geographic distribution. Late Miocene (Huayquerian) Argentina and possible Venezuela.

Paleoecology. *Stylocynus paranensis* was a large sparassodont (35.3 kg or 26.8 kg based on tooth equations; Wroe et al. 2004a; Prevosti et al. 2013, respectively), and probably with omnivorous/mesocarnivorous diet (Prevosti et al. 2013).

Comments. Material assigned to *S. paranensis* includes fragments of maxillae and dentaries with associated teeth from the Mesopotamian area in Argentina (e.g., Marshall 1979). Material tentatively assigned to *Stylocynus* has been recovered from northwestern Argentina (Babot and Ortiz 2008) and Urumaco in Venezuela (Linares 2004). The latter material appears to have been misplaced (Sánchez-Villagra per. com. 2015).

Marshall (1979) considered *S. paranensis* as a highly specialized borhyaenoid. Alternatively, cladistic analyses recovered *S. paranensis* outside major sparassodont groups (Forasiepi 2009; Engelman and Croft 2014; Forasiepi et al. 2015; Fig. 3.2), and in contrast, its tooth morphology could interpreted as plesiomorphic.

Hondadelphidae Marshall et al. 1990
Hondadelphys Marshall 1976a
(Tables 3.1 and 3.2)

Included species. *Hondadelphys fieldsi* Marshall 1976a.

Temporal and geographic distribution. Middle Miocene (Laventan), Colombia.

Paleoecology. Small size (3.7 kg estimated with tooth equations) and omnivorous (Prevosti et al. 2013).

Comments. *Hondadelphys fieldsi* has provoked different opinions regarding its affinities. Marshall (1976a) originally considered *H. fieldsi* to be a didelphid later placing it in Sparassodonta (Marshall et al. 1990; Goin 1997). Goin (1997) suggested that *H. fieldsi* was probably related to thylacosmilids, but cladistic analysis places outside of any major group (Forasiepi 2009; see also Marshall et al. 1990; Fig. 3.2). This hypothesis is in agreement with the "museum" model for the tropics, as a region in which old lineages have been able to persist for longer than in higher latitudes (see also Suarez et al. 2015).

Hathliacynidae Ameghino 1894

Hathliacynidae are small- to medium-sized sparassodonts, with long and slender skulls and dentaries that are fox- or weasel-like in appearance (Figs. 3.4, 3.5 and 3.6). All hathliacynids may have had a hypercarnivorous diet as suggested by the RGA dental index; however, the molars have very reduced crushing surfaces, which suggest the possibility of some dietary flexibility.

Hathliacynidae is a monophyletic group and includes the common ancestor of *Sipalocyon* and *Cladosictis* and all the taxa with a closer relationship to them than to borhyaenoids (Forasiepi 2009; Fig. 3.2). The oldest hathliacynid is from the late Oligocene (Deseadan) and the youngest, *Borhyaenidium riggsi*, is from the mid-Pliocene (Chapadmalalan). Putative hathliacynids have been claimed for the middle Eocene (Casamayoran; Marshall 1981) or late Oligocene (La Cancha association; Goin et al. 2010), which may considerably extend the stratigraphic range of the group. However, systematic interpretation of these putative hathliacynids requires the support of a phylogenetic analysis.

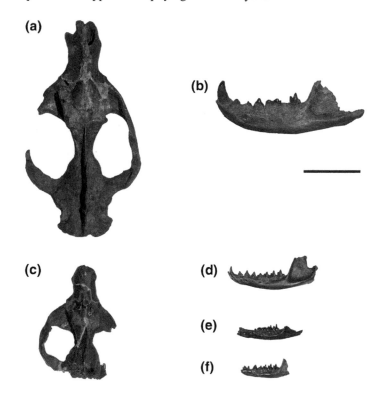

Fig. 3.4 Hathliacynidae from the early Miocene (Santacrucian Age), Santa Cruz Formation, Patagonia. *Cladosictis patagonica* (MACN-A 5927–5929, single specimen) cranium in dorsal view (**a**) and left dentary in lateral view (**b**); *Sipalocyon gracilis* (YPM PU 15154) nearly complete skull in dorsal view (**c**) and (YPM PU 15373) left dentary in lateral view (**d**); *Pseudonotictis pusillus* (MLP 11–26) left dentary in lateral view; *Perathereutes pungens* (MACN-A 684, Holotype) left dentary in lateral view (**e**). Scale bar = 5 cm

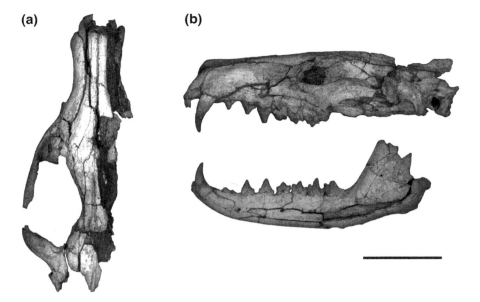

Fig. 3.5 *Acyon myctoderos* (MNHN-Bol-V-003668, Holotype), from the middle Miocene (Laventan Age), Quebrada Honda (Bolivia), cranium in dorsal view (**a**) and cranium with left dentary in lateral view (**b**). Scale bar = 5 cm

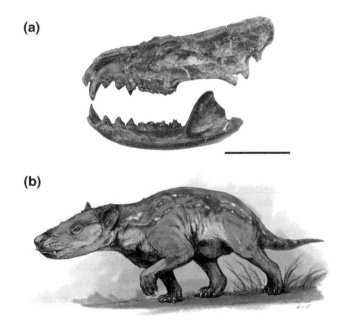

Fig. 3.6 *Cladosictis patagonica* (YPM PU 15046) from the early Miocene (Santacrucian Age), 15 km south of Coy Inlet, Patagonia (Argentina), cranium and left dentary in lateral view (**a**); artistic reconstruction by Jorge Blanco (**b**). Scale bar = 5 cm

The phylogenetic analyses are not conclusive on the affinities of *Cladosictis*. Some cladistic hypotheses recovered *Cladosictis* and *Acyon* as sister taxa (Forasiepi et al. 2006; Engelman and Croft 2014), or alternatively *Sallacyon hoffstetteri* (Engelman and Croft 2014), *Notogale mitis* (de Muizon 1999; Babot et al. 2002), or these taxa plus *Sipalocyon* (Suarez et al. 2015) as sister taxa.

Contrerascynus Mones 2014
(Tables 3.1 and 3.2)

Included species. *Contrerascynus borhyaenoides* (Contreras 1990).
Temporal and geographic distribution. Late Miocene (Chasicoan), Argentina.
Paleoecology. Based on the length of m3 (Gordon 2003, using dasyurids), estimations suggested a body mass of 12.6 kg. The calculation of the RGA suggested hypercarnivorous diet.
Comments. *Contrerascynus borhyaenoides* is known only by a fragmentary dentary with a dental morphology that resembles *Acyon* (=*Anatherium* in Contreras 1990); however, it is larger and comparable to *Lycopsis viverensis* from the Chasicoan of Pampean area.
The finding of the material in the Quebrada de Ullúm, San Juan (Argentina), provides a new area for prospecting in the central-western part of Argentina.

Notictis Ameghino 1889
(Tables 3.1 and 3.2)

Included species. *Notictis ortizi* Ameghino 1889.
Temporal and geographic distribution. Late Miocene (Huayquerian), Argentina.
Paleoecology. The body mass of *N. ortizi* was estimated at 0.9 kg, and the diet was considered hypercarnivorous (Prevosti et al. 2013; Zimicz 2014).
Comments. *Notictis ortizi* is represented only by the holotype, a fragment of dentary with partial dentition. This taxon is the smallest Huayquerian sparassodont, similar in size to the Santacrucian *Pseudonotictis pusillus*. Phylogenetic affinities between these taxa were suggested by Marshall (1981). This hypothesis awaits testing.

Notocynus Mercerat 1891
(Tables 3.1 and 3.2)

Included species. *Notocynus hermosicus* Mercerat 1891.
Temporal and geographic distribution. Early Pliocene (Montehermosan), Argentina.
Paleoecology. *Notocynus hermosicus* was a small hathliacynid with a body mass estimation of 1.77 kg or 3.2 kg (Prevosti et al. 2013; Wroe et al. 2004a, respectively). The diet was considered hypercarnivorous (Prevosti et al. 2013; Zimicz 2014).
Comments. *Notocynus hermosicus* is known only by its holotype, a single fragmentary dentary, collected from Monte Hermoso type locality (Cabrera 1927; Marshall 1981). *Notocynus hermosicus* and *Thylacosmilus atrox* are currently the only two Montehermosan sparassodonts known.

Notogale Loomis 1914
(Tables 3.1 and 3.2)

Included species. *Notogale mitis* (Ameghino 1897).

Temporal and geographic distribution. Late Oligocene (Deseadan), Argentina and Bolivia.

Paleoecology. *Notogale mitis* was similar in size to *Cladosictis* according to linear measurements of the dentition (e.g., Marshall 1981). However, tooth equations considerably understimated its probable body size (2.72 kg Prevosti et al. 2013; 3.4 kg in Zimicz 2012 vs. 6.6 kg for *Cladosictis patagonica* in Ercoli and Prevosti 2011). A hypercarnivorous diet was suggested on the basis of RGA tooth equations (Zimicz 2012; Prevosti et al. 2013).

Comments. *Notogale mitis* has a broad distribution in South America, from Patagonia in the south to the Bolivian locality of Salla in the north, where it is the most abundant sparassodont (sensu Marshall 1981). Marshall (1981) suggested that *Notogale* was closely related to *Cladosictis*. This hypothesis was supported by de Muizon (1999) and Babot et al. (2002) with a cladistic approach placing them as sister taxa. Other studies have recovered *Notogale* and *Sipalocyon* or *Sallacyon* as sister taxa (Forasiepi 2009; Engelman and Croft 2014; Suarez et al. 2015).

Perathereutes Ameghino 1891
(Fig. 3.4e; Tables 3.1 and 3.2)

Included species. *Perathereutes pungens* Ameghino 1891.
Temporal and geographic distribution. Early Miocene (Santacrucian), Argentina.
Paleoecology. *Perathereutes pungens* was intermediate in size between *Sipalocyon gracilis* and *Pseudonotictis pusillus*. On the basis of tooth equations, Wroe et al. (2004a) provided values of 2.5 kg, whereas Prevosti et al. (2013) and Zimicz (2014) suggested ∼1 kg. The diet was suggested as hypercarnivorous (Prevosti et al. 2013).

Comments. This taxon is known from scarce material from Patagonia. Marshall (1981) suggested phylogenetic affinities between *P. pungens* and the late Miocene *Borhyaenidium musteloides*. This hypothesis awaits testing.

Pseudonotictis Marshall 1981
(Fig. 3.4f; Tables 3.1 and 3.2)

Included species. *Pseudonotictis pusillus* (Ameghino 1891) (type species) and *P. chubutensis* Martin and Tejedor 2007.
Temporal and geographic distribution. *Pseudonotictis pusillus* comes from the early Miocene (Santacrucian), Argentina; *P. chubutensis* from the middle Miocene (Colloncuran), Argentina.
Paleoecology. *Pseudonotictis pusillus* and *P. chubutensis* are the smallest sparassodonts from the Santacrucian and Colloncuran outcrops, respectively. They would have had a size similar to the extant weasel (*Mustela frenata*). The body mass of *P. pusillus* was estimated at 1.17 kg based on the centroid size of the humerus (Ercoli and Prevosti 2011), roughly similar to the values obtained with tooth variables (0.93 kg in Prevosti et al. 2013). The body mass of *P. chubutensis* provided smaller values (0.89 kg in Zimicz 2014).

Study of locomotion in *P. pusillus* suggested scansorial habits with evident arboreal capabilities (Argot 2003a; Ercoli et al. 2012). A hypercarnivorous diet was suggested for *P. pusillus* on the basis of RGA tooth equations (Prevosti et al. 2013).

Comments. The genus was erected by Marshall (1981) to include *P. pusillus*, which closely resembles the late Miocene (Huayquerian) *Notictis ortizi*.

Sallacyon Villarroel and Marshall 1982
(Tables 3.1 and 3.2)

Included species. *Sallacyon hoffstetteri* Villarroel and Marshall 1982.
Temporal and geographic distribution. Late Oligocene (Deseadan), Bolivia.
Paleoecology. *Sallacyon hoffstetteri* was slightly smaller than *Sipalocyon gracilis* according to Villarroel and Marshall (1982). Body mass estimations provided values similar to or sligthly larger than *Sipalocyon gracilis* (1.13 kg in Prevosti et al. 2013; 3 kg in Zimicz 2012). The diet was considered hypercarnivorous (Prevosti et al. 2013; Zimicz 2012).

Comments. *Sallacyon hoffstetteri* is a Neotropical sparassodont hitherto known only from the Deseadan outcrops of Bolivia (Villarroel and Marshall 1982; Petter and Hoffstetter 1983; de Muizon 1999). Its dental morphology and dentary resemble the early Miocene (Santacrucian) *Perathereutes pungens* and *Sipalocyon gracilis*, to whom it may be phylogenetically close (Villarroel and Marshall 1982; Petter and Hoffstetter 1983). Alternatively, cladistic studies positioned this taxon together with *Notogale* (Forasiepi 2009; Suarez et al. 2015), or in a basal branch among hathliacynids (de Muizon 1999; Babot 2005; Forasiepi et al. 2006).

Sipalocyon Ameghino 1887
(Fig. 3.4c, d; Tables 3.1 and 3.2)

Included species. *Sipalocyon gracilis* Ameghino 1887 (type species), *S. externa* Ameghino 1902b, and *S. obusta* (Ameghino 1891) (nomen dubium).

Temporal and geographic distribution. *Sipalocyon externa* comes from the early Miocene (Colhuehuapian), Argentina; *S. obusta* from the early Miocene (Santacrucian), Argentina; *S. gracilis* from the early to middle Miocene (Santacrucian and Friasian), Argentina and Chile.

Paleoecology. On the basis of the centroid size of the ulna, the body mass of *S. gracilis* was estimated to be 2.11 kg (Ercoli and Prevosti 2011). This value is in the range of the predictions of Argot (2003a) (between 1 kg and 5 kg according to different specimens) based on postcranial equations, and the estimations of Vizcaíno et al. (2010; 1.93 kg), Prevosti et al. (2013; 1.96 kg), and Zimicz (2014; 3.15 kg) based on tooth equations. The other two species are in the same range of size (*S. externa*: 0.93 kg or 2.48 kg; *S. obusta*: 1.83 kg or 2.81 kg based on tooth equations; Prevosti et al. 2013; Zimicz 2014, respectively).

Originally, Sinclair (1906) suggested arboreal habits for *S. gracilis*; however, recent studies suggested scansorial locomotion (Marshall 1978; Argot 2003a, 2004a; Ercoli et al. 2012). The limbs would have had skillful manipulative capabilities (Argot 2003a, 2004a). A hypercarnivorous diet was suggested for *S. gracilis* and *S. obusta* on the basis of RGA tooth equations (Prevosti et al. 2013; Zimicz 2014).

Comments. *Sipalocyon gracilis* is a common taxon in the Santacrucian levels of Patagonia. On the contrary, the second Santacrucian species, *S. obusta,* is extremely scarce and is "virtually identical to …*S. gracilis*" (Marshall 1981: 60), but with shallower and more slender dentary and m4 with more reduced talonid. These differences could represent intraspecific variability in *S. gracilis.* Consequently, *S. obusta* is only tentatively recognized and considered nomen dubium.

Marshall (1981) suggested that *Sipalocyon* was phylogenetically close to *Perathereutes.* Alternatively, cladistic analyses recovered *Sipalocyon* as the sister taxon of *Notogale* plus *Cladosictis* (de Muizon 1999; Babot et al. 2002); sister taxon of *Notogale* (Forasiepi et al. 2006; Engelman and Croft 2014), or *Notogale* plus *Sallacyon* (Forasiepi 2009; Engelman and Croft 2014; Forasiepi et al 2015; Suarez et al. 2015).

?Hathliacynidae
Procladosictis Ameghino 1902a
(Tables 3.1 and 3.2)

Included species. *Procladosictis anomala* Ameghino 1902a.
Temporal and geographic distribution. Latest middle–earliest late Eocene (Mustersan), Argentina.
Paleoecology. The body mass of *P. anomala* was estimated at 8.9 kg (Zimicz 2012) with a possible hypercarnivorous diet (Prevosti et al. 2013).
Comments. *Procladosictis anomala* is known only by its type specimen, a fragment of maxilla with dentition. The molars have broad stylar shelves and deep ectoflexus, which are unusual features among sparassodonts. Marshall (1981) considered *P. anomala* a hathliacynid; however, new material is needed to illuminate its relationships (Forasiepi 2009).

Borhyaenoidea Simpson 1930

Borhyaenoidea includes medium- to large-sized sparassodonts, recorded from the middle Eocene (Casamayoran) to the latest early Pliocene (Chapadmalalan) (Figs. 3.7, 3.8, 3.9, 3.10, 3.11, 3.12 and 3.13). It includes the common ancestor of *Prothylacynus* and *Borhyaena* and all the taxa that are more closely related to them than to hathliacynids (Forasiepi 2009). There is significant morphological disparity in the cranium, dentition, and postcranial skeleton. Some taxa were slender and light, such as the fox-like *Lycopsis viverensis,* others were massive and robust, exhibiting deep dentaries sometimes fused at symphysis, such as the bear-like *Arctodictis munizi,* while others had hypertrophied sabertooth canines, such as *Thylacosmilus.* The diet as indicated by the RGA dental index was hypercarnivorous for most of the group; however, some borhyaenoids with larger protocones and broader talonids may have had a more flexible diet than proborhyaenids, thylacosmilids, and borhyaenids, whose molars lack crushing surfaces.

Dukecynus Goin 1997
(Tables 3.1 and 3.2)

Included species. *Dukecynus magnus* Goin 1997.
Temporal and geographic distribution. Middle Miocene (Laventan), Colombia.

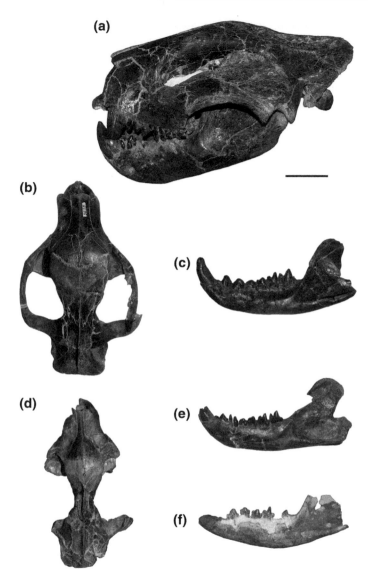

Fig. 3.7 Borhyaenoidea from the early Miocene (Santacrucian Age), Santa Cruz Formation, Patagonia, Argentina. *Arctodictis munizi* (CORD-PZ 1210) cranium and left dentary in lateral view (**a**); *Borhyaena tuberata* (MPM-PV 3625) cranium in dorsal view (**b**) and (MACN-A 12700) inverted right dentary in lateral view (**c**); *Prothylacynus patagonicus* (MACN-A 5931–5937, single specimen) cranium of juvenile specimen (**d**), (MACN-A 706–720, Holotype) left dentary in lateral view (**e**); *Lycopsis torresi* (MLP 11-113, Holotype), left dentary in lateral view (**f**). Scale bar = 5 cm

Paleoecology. The body mass of *D. magnus* was estimated at 68.4 kg or 52.6 kg (Wroe et al. 2004a; see also comment below) or 24.6 kg (Prevosti et al. 2013) on the basis of dental variables. The hypercarnivorous diet was inferred the same as for *Lycopsis torresi* (Prevosti et al. 2013).

Comments. *Dukecynus magnus* was originally considered a member of the paraphyletic "prothylacynines" and related to the Chasicoan *Pseudolycopsis* (Goin 1997). Cladistic reconstructions position most of the "prothylacynines" (i.e., *Prothylacynus, Lycopsis*) among basal clades of Borhyaenoidea, a likely position also for *D. magnus*.

In his monograph about borhyaenoids, Marshall (1978) mentioned a putative *Arctodictis* specimen (UCMP 39250). This material was later assigned to *Dukecynus magnus* (Goin 1997; Forasiepi et al. 2004). For this specimen, Wroe et al. (2004a) obtained a body mass of 51.6 kg.

Lycopsis Cabrera 1927
(Figs. 3.7f, 3.8; Tables 3.1 and 3.2)

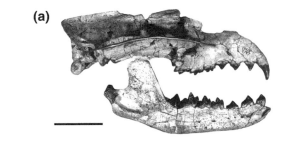

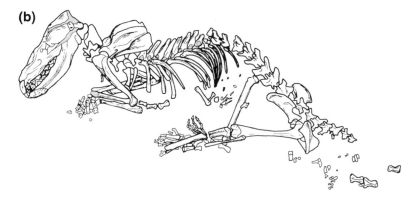

Fig. 3.8 *Lycopsis longirostrus* (UCMP 38061, Holotype), from the middle Miocene (Laventan Age), La Venta (Colombia), cranium and right dentary in lateral view (**a**) and line drawing of the skeleton (taken from Marshall 1977b) (**b**). Scale bar = 5 cm

Included species. *Lycopsis torresi* Cabrera 1927 (type species), *L. longirostrus* Marshall 1977b, *L. viverensis* Forasiepi et al. 2003, *L. padillai* Suarez, Forasiepi, Goin, Jaramillo 2015.

Temporal and geographic distribution. *Lycopsis torresi* was found in the early Miocene (Santacrucian), Argentina; *L. padillai* and *L. longirostrus* from middle Miocene (Colloncuran and Laventan, respectively), Colombia; *L. viverensis* from late Miocene (Chasicoan), Argentina.

Paleoecology. The four species are clearly differentiated by size. A recent study (Suarez et al. 2015) on the basis of the upper molar occlusal row length predicted a body mass for *Lycopsis torresi* at ~ 27 kg, and in the range of *L. padillai* with ~ 22 kg. The smallest species *L. viverensis* was estimated at ~ 18 kg while the largest *L. longirostrus* at ~ 44 kg. These results are roughly consistent with Prevosti et al. (2013) based on tooth measurements (*L. torresi*: 31.5 kg; *L. viverensis*, 11 kg; *L. longirostrus*: 42.5 kg), but are overestimated compared with the values obtained using postcranial variables. On the basis of the centroid size of the ulna and humerus, Ercoli and Prevosti (2011) provided values of 29.77 kg for *L. longirostrus*, while on the basis of postcranial variables Argot (2004b) suggested 17.1 kg and Wroe et al. (2004a) calculated 12.8 kg for the same taxon. For *L. torresi* and on the basis of tooth variables, Wroe et al. (2004a) obtained 19.4 kg.

The only species of *Lycopsis* with a known postcranium is *L. longirostrus*. For this taxon, Argot (2004a, b) and Ercoli et al. (2012) concurred in viewing this taxon as possessing terrestrial progression and reduced running capabilities. The limbs were plantigrade with grasping abilities in the forelimb.

The diet was hypercarnivorous, as suggested by the RGA dental equations (Prevosti et al. 2013). In the abdominal area of *L. longirostrus,* "between the ribs and right tibia were broken rodent bones and an upper molar of *Scleromys colombianus*" (Caviomorpha), (Marshall 1977b: 641)—evidently, the predator's last meal.

Comments. The holotype of *L. longirostrus* consists of an almost complete skeleton, still articulated (Marshall 1977b; Fig. 3.8). The last molar is not completely erupted (Forasiepi and Sánchez-Villagra 2014), which suggests the specimen was a subadult.

The monophyly of the genus *Lycopsis* has been recently supported (Suarez et al. 2015), with the Laventan *L. longirostrus* recorded as the most basal member of the genus. The results of Suarez et al. (2015) implied a diversification within (or before) the early Miocene and that *Lycopsis* had at least two migration events between the Neotropical and the temperate regions, with divergence into new species in the temperate region. The persistence of a basal taxon (*L. longirostrus*) in Neotropical areas in younger (middle Miocene) ages was interpreted under the "museum" evolutionary model (Suarez et al. 2015).

Lycopsis is the genus of Sparassodonta with the broadest distribution, ranging from La Guajira (Colombia) to Patagonia (Argentina) (Suarez et al. 2015).

Pharsophorus Ameghino 1897
(Tables 3.1 and 3.2)

Included species. *Pharsophorus lacerans* Ameghino 1897 (type species) and
P. tenax Ameghino 1897.

Temporal and geographic distribution. *Pharsophorus lacerans* and *P. tenax*
come from the late Oligocene (Deseadan), Argentina (both species) and Bolivia (the
former only).

Paleoecology. The two *Pharsophorus* species differ clearly by size. Estimations
for *P. lacerans* are ~27 kg, with hypercarnivorous diet (Zimicz 2012; Prevosti
et al. 2013). Estimations for *P. tenax* have suggested 18.7 kg (Zimicz 2012) and
mesocarnivorous diet (Zimicz 2012).

Comments. Of the two species, *P. lacerans* is the largest and the best represented. *Pharsophorus tenax* was named by Ameghino (1897) on the basis of an
isolated m1. The holotype was missing, and at the time of Marshall's revision of the
group (1978), a neotype was designated. Simultaneously, Patterson and Marshall
(1978) synonymized *P. tenax* with *P. lacerans*. In 1997, the material used by
Ameghino to define the species was found by A. Ramos in MACN collections and
given the number MACN-A 11113. A new revision of this species is required to
restore the original holotype and to evaluate the validity of the taxon. Provisionally,
we consider here *P. tenax* as a valid species, as Marshall (1978) had concluded.

Originally, Ameghino (1897) and later Marshall (1978) suggested affinities
between *Pharsophorus* and *Borhyaena*, while Patterson and Marshall (1978) suggested possible phylogenetic relationships with thylacosmilids. In agreement with
both views, cladistic analyses recovered *P. lacerans* as the sister taxon of thylacosmilids, proborhyaenids, and borhyaenids (Forasiepi 2009; Engelman and Croft
2014; Forasiepi et al. 2015; Suarez et al. 2015).

Plesiofelis Roth 1903
(Tables 3.1 and 3.2)

Included species. *Plesiofelis schlosseri* Roth 1903.

Temporal and geographic distribution. Latest middle–earliest late Eocene
(Mustersan), Argentina.

Paleoecology. *Plesiofelis schlosseri* was slightly larger than *Pharsophorus lacerans*. Body mass estimations suggested ~45 kg and ~ 32 kg (Zimicz 2012 and
Prevosti et al. 2013, respectively) with hypercarnivorous diet.

Comments. Cabrera (1927) and Simpson (1948) considered *Plesiofelis* to be
synonym of *Pharsophorus*. Later, Marshall (1978) recognized *Plesiofelis* as a valid
taxon. As suggested by their close phylogenetic affinity, the two taxa have similar
tooth morphology (Marshall 1978; Forasiepi et al. 2015; an alternative interpretation was presented by Goin et al. 2007).

Prothylacynus Ameghino 1891
(Fig. 3.7d, e; Tables 3.1 and 3.2)

Included species. *Prothylacynus patagonicus* Ameghino 1891.

Temporal and geographic distribution. Early to middle Miocene (Santacrucian,
Friasian, and Colloncuran), Argentina and Chile.

Paleoecology. For *P. patagonicus*, estimations on the basis of the centroid size of the ulna and tibia suggested a body mass of 31.8 kg (Ercoli and Prevosti 2011), similar to ~30 kg of Argot (2003b) on the basis of postcranial variables, and comparable to a wolverine (*Gulo gulo*). Other predictions are somewhat different (e.g., 26.8 kg using variables from the femur in Wroe et al. 2004a; 13.83 kg and 20.6 kg using dental measurements in Vizcaíno et al. 2010; Prevosti et al. 2013, respectively).

Originally, Sinclair (1906) and Marshall (1978) indicated terrestrial locomotion for *P. patagonicus*; however, more recent analysis has suggested scansorial adaptations (Argot 2003b, 2004a; Ercoli et al. 2012). The limb architecture suggested plantigrade (Sinclair 1906) or semiplantigrade posture (Argot 2003b, 2004a), with skillful manipulative behavior. *Prothylacynus patagonicus* was possibly a more active predator than contemporaneous *Borhyaena tuberata*, with a flexible vertebral column that allowed powerful jumps from a crouched position (Argot 2003b, 2004a). A hypercarnivorous diet was estimated based on the RGA dental index (Prevosti et al. 2013).

Comments. *Prothylacynus patagonicus* is a species frequently found in Santacrucian outcrops, which have yielded both cranial and postcranial material (Sinclair 1906).

Traditionally, *Prothylacynus* was grouped with *Lycopsis*, *Pseudolycopsis*, *Pseudothylacynus*, *Stylocynus*, and *Dukecynus* in the subfamily Prothylacyninae (Marshall 1979; Marshall et al. 1990; Goin 1997). Later analyses considered Prothylacyninae paraphyletic with *Prothylacynus* placed among basal borhyaenoids (Babot 2005; Forasiepi et al. 2006, 2015; Forasiepi 2009; Engelman and Croft 2014; Suarez et al. 2015).

Pseudolycopsis Marshall 1976b
(Tables 3.1 and 3.2)

Included species. *Pseudolycopsis cabrerai* Marshall 1976b.
Temporal and geographic distribution. Late Miocene (Chasicoan), Argentina.
Paleoecology. Using dental variables, the body mass of *P. cabrerai* was suggested to be 24 kg or 14.4 kg (Wroe et al. 2004a; Prevosti et al. 2013, respectively). Diet similar to *Lycopsis* species (Prevosti et al. 2013).

Comments. *Pseudolycopsis cabrerai* is known by a fragment of palate (Marshall 1976b).

Marshall suggested that *Pseudolycopsis* was likely related to the genus *Lycopsis* (Marshall 1976b, 1979).

Pseudothylacynus Ameghino 1902b
(Tables 3.1 and 3.2)

Included species. *Pseudothylacynus rectus* Ameghino 1902b.
Temporal and geographic distribution. Early Miocene (Colhuehuapian), Argentina.
Paleoecology. Body mass estimates using dental variables are 14 kg and 19.7 kg (Wroe et al. 2004a; Prevosti et al. 2013, respectively). Diet was hypercarnivorous.

Comments. Little material is identified from this species. The anatomy of the dentition closely resembles that to *Prothylacynus*, suggesting close phylogenetic affinities (Marshall 1979).

Proborhyaenidae Ameghino 1897

Traditionally, Proborhyaenidae was considered the group that includes the largest hypercarnivorous mammalian predators from the middle Eocene (Casamayoran) to the late Oligocene (Deseadan) of South America (Marshall 1978; Bond and Pascual 1983; Petter and Hoffstetter 1983; Babot et al. 2002; Fig. 3.9). However, recent cladistic analyses are not congruent regarding their monophyly. Some studies have concluded that they are monophyletic (e.g., Babot et al. 2002; Engelman and Croft 2014), although the most exhaustive analysis that included several species of this group recorded them as paraphyletic (Babot 2005; Argot and Babot 2011).

Arminiheringia Ameghino 1902a
(Tables 3.1 and 3.2)

Included species. *Arminiheringia auceta* Ameghino 1902a (type species), *A. cultrata* Ameghino 1902a, and *A. contigua* Ameghino 1904.

Temporal and geographic distribution. Middle Eocene (Casamayoran), Argentina.

Paleoecology. *Arminiheringia* included large-size sparassodonts: *A. auceta*, 31.3 kg and 31.7 kg (Prevosti et al. 2013; Zimicz 2012, respectively); *A. cultrata*, 24 kg and 25.7 kg (Prevosti et al. 2013; Zimicz 2012, respectively), and *A. contigua*, 18.5 kg (Zimicz 2012). The diet was hypercanivorous, with capacity to break bones (Zimicz 2012).

Comments. Three valid species of *Arminiheringia* are considered following Babot (2005). However, the validity of some of these taxa has been questioned, because of a lack of diagnostic features. Simpson (1948) considered *A. cultrata* to be a synonym of *A. contigua*, but Marshall (1978) concluded *A. cultrata* was a synonym of *A. auceta*.

Classically, *Arminiheringia* was considered phylogenetically close to *Thylacosmilus* (Scott 1937), but a later diagnosis did not confirm that their morphological resemblances indicated close ancestry (Simpson 1948; Marshall 1976c, 1978). Recent cladistic analysis recovered *A. auceta* and *Callistoe vincei* as sister taxa (Babot 2005).

Callistoe Babot et al. 2002
(Tables 3.1 and 3.2)

Included species. *Callistoe vincei* Babot et al. 2002.

Temporal and geographic distribution. Middle Eocene (Vacan subage of the Casamayoran), Argentina.

Paleoecology. The body mass of *C. vincei* was estimated using postcranial variables, producing a value of ~23 kg (Argot and Babot 2011), in the range of *Thylacinus cynocephalus*. Estimations using tooth variables instead indicated body masses of 32.6 kg and 27.75 kg (Argot and Babot 2011; Prevosti et al. 2013, respectively).

Terrestrial locomotion was suggested for *C. vincei* with limbs that favored flexion/extension parasagittal movements, rather than pronation/supination (Argot and Babot 2011). The long claws in the forelimb suggested the capacity to dig for small prey in burrows, while the mobile thumb suggested that it could grab and manipulate objects (Argot and Babot 2011).

Comments. *Callistoe vincei* is known by exceptional material, represented by a cranium and complete postcranial elements (Babot et al. 2002; Babot 2005; Argot and Babot 2011). The holotype of *C. vincei* is the best preserved Paleogene sparassodont recovered to date (2016).

Callistoe vincei has been found in Pampa Grande, Salta, northern Argentina (Babot et al. 2002). Paleoenvironmental reconstructions suggested it lived in a temperate humid forest biome (Powell et al. 2011). *Callistoe vincei* is large and slender compared to other proborhyaenids. Cladistic reconstructions recovered *C. vincei* the sister taxon of *Arminiheringia* (Babot 2005) or alternatively, as sister taxon of *Paraborhyaena boliviana* (Babot et al. 2002; Engelman and Croft 2014).

Paraborhyaena Hoffstetter and Petter 1983
(Tables 3.1 and 3.2)

Included species. *Paraborhyaena boliviana* Hoffstetter and Petter 1983.
Temporal and geographic distribution. Late Oligocene (Deseadan), Bolivia.
Paleoecology. The body mass should be in the range of *A, auceta*, considering the similarity of its measurements. Diet was hypercarnivorous (Prevosti et al. 2013).

Comments. *Paraborhyaena boliviana* is a Neotropical sparassodont known by a single specimen (Hoffstetter and Petter 1983). Originally, Petter and Hoffstetter (1983) considered close affinities between *P. boliviana* and *Arminiheringia* and *Proborhyaena* species. Later, cladistic analysis alternatively grouped *P. boliviana* and *C. vincei* (Babot et al. 2002; Engelman and Croft 2014), *P. boliviana* and *Proborhyaena gigantea* plus thylacosmilids (Babot 2005), or *P. boliviana* and thylacosmilids (Suarez et al. 2015).

Proborhyaena Ameghino 1897
(Fig. 3.9; Tables 3.1 and 3.2)

Fig. 3.9 *Proborhyaena gigantea* (AMNH 29576), from the late Oligocene (Deseadan), Rinconada de López, Patagonia (Argentina), right dentary in lateral view. Scale bar = 5 cm

Included species. *Proborhyaena gigantea* Ameghino 1897.

Temporal and geographic distribution. Late Oligocene (Deseadan), Argentina and Uruguay.

Paleoecology. *Proborhyaena gigantea* is the largest known sparassodont, with a size similar to the South American spectacled bear (*Tremarctos ornatus*). Body mass estimations are disparate. Zimicz (2012) and Prevosti et al. (2013) provided values between 93 kg and 153.6 kg, respectively. Sorkin (2008) based on the lower canine–last molar length suggested 600 kg, a value that is surely overestimated. *Proborhyaena gigantea* was a hypercarnivorous bone-cracker, and likely able to actively predate in the fashion of living hyenas (*Crocuta crocuta*) (Zimicz 2012).

Proborhyaena gigantea may have had a powerful canine bite as suggested by mandibular force profiles (Blanco et al. 2011), with unpredictable direction of forces and capabilities to break bones at the level of the last molar.

Comments. *Proborhyaena gigantea* had a large distribution through the southern cone of South American with findings in Patagonia, Mendoza, and Uruguay (Mones and Ubilla 1978; Patterson and Marshall 1978; Bond and Pascual 1983).

Originally, Marshall (1978) suggested close affinities between *P. gigantea* and *Arminiheringia*. Alternatively, cladistic analysis recovered *P. gigantea* as the sister taxon of thylacosmilids (Babot 2005).

Thylacosmilidae Riggs 1933

Thylacosmilidae includes taxa with the most unusual morphology among South American native predators. The overall cranial morphology resembles sabertooth felids (Chap. 4) in that both acquired large hypertrophied upper canines (Figs. 3.10 and 3.11). This is renowned as a classic example of convergent evolution (e.g., Riggs 1933, 1934; Simpson 1971; Marshall 1976c, 1977a; Turnbull 1978; Turnbull and Segall 1984; Churcher 1985). Other features include short cranium with massive snout, and presence of bony auditory bulla; mandibles with a subvertical symphyseal flange, shallow ramus with the alveolar and ventral edges subparallel and straight; small but deep masseteric fossa, poorly inflected angle, low condyle in relation to the alveolar plane; cheekteeth bowed and simplified molar structures favoring shearing and suggesting a highly specialized hypercarnivorous diet (Marshall 1976c; Goin and Pascual 1987; Goin 1997; Mones and Rinderknecht 2004; Forasiepi and Carlini 2010).

Thylacosmilidae is a monophyletic group and includes three species: *Anachlysictis gracilis, Patagosmilus goini,* and *Thylacosmilus atrox*. Its stratigraphic range dates from the middle Miocene (Colloncuran) to the latest early Pliocene (Chapadmalalan). In addition, a putative thylacosmilid represented by an isolated upper molar has been collected from the early Miocene (Colhuehuapian) of Patagonia (Goin et al. 2007). If the assignation of the Patagonian specimen is correct, the stratigraphic range of the group dates back 5 Ma more than currently accepted (Goin et al. 2007). Another putative thylacosmilid is represented by a middle Miocene (Laventan) specimen (Goin 1997). It has a much more generalized morphology than other thylacosmilids, with the symphyseal area of the dentary, the morphology of the maxilla, and the general structure of the dentition recalling thylacosmilids. This Laventan taxon could either represent a stem or basal thylacosmilid, or alternatively a different sparassodont lineage with incipient and convergent sabertooth architecture (Goin 1997).

Anachlysictis Goin 1997
(Fig. 3.10a; Tables 3.1 and 3.2)

Included species. *Anachlysictis gracilis* Goin 1997.
Temporal and geographic distribution. Middle Miocene (Laventan), Colombia.
Paleoecology. The body mass was estimated with tooth equations in ~ 18 and 16 kg (Wroe et al. 2004a; Prevosti et al. 2013, respectively).
Comments. *Anachlysictis gracilis* is known only by its holotype, a mandible with dentition and associated postcranium, which exhibit several plesiomorphies compared to *Thylacosmilus atrox* (Goin 1997; Forasiepi and Carlini 2010).

Patagosmilus Forasiepi and Carlini 2010
(Fig. 3.10b, c; Tables 3.1 and 3.2)

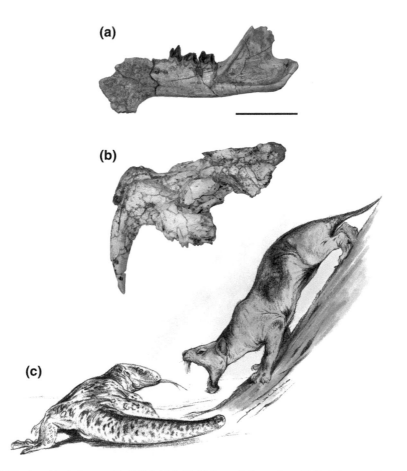

(a)

(b)

(c)

Fig. 3.10 *Anachlysictis gracilis* (IGM 184247, Holotype) from the middle Miocene (Laventan Age), La Venta (Colombia), inverted right dentary in lateral view. *Patagosmilus goini* (MLP 07-VII-1-1), partial cranium in lateral view (**b**); artistic reconstruction by Jorge Blanco (**c**). Scale bar = 5 cm

Included species. *Patagosmilus goini* Forasiepi and Carlini 2010.

Temporal and geographic distribution. Middle Miocene (Colloncuran) Argentina.

Paleoecology. In view of its linear measurements, *P. goini* probably had a body mass similar to that of *Anachlysictis gracilis* (Prevosti et al. 2013). Diet was hypercarnivorous.

Comments. *Patagosmilus goini* is represented by a partial skull and a few postcranial elements. From the middle Miocene (Colloncuran) of Patagonia, this is to date the oldest definite thylacosmilid. Cladistic analyses recovered *P. goini* and *Thylacosmilus atrox* as sister taxa (Engelman and Croft 2014; Forasiepi et al. 2015; Suarez et al. 2015).

Thylacosmilus Riggs 1933
(Fig. 3.11; Tables 3.1 and 3.2)

Fig. 3.11 *Thylacosmilus atrox* (MMP 1443), from the late early Pliocene (Chapadmalalan Age), Chapadmalal (Argentina), complete cranium and dentary in lateral view (**a**); artistic reconstruction by Jorge Blanco (**b**). Scale bar = 5 cm

(a)

(b)

Included species. *Thylacosmilus atrox* Riggs 1933.

Temporal and geographic distribution. Late Miocene to latest early Pliocene (Huayquerian, Montehermosan, and Chapadmalalan), Argentina and Uruguay.

Paleoecology. The efforts to calculate the body mass of *T. atrox* have yielded substantialy different values, which could be either a response to marked intraspecific variability, different methodologies for the estimations, or both. Based on the centroid size of the tibia and ulna, Ercoli and Prevosti (2011) obtained 117.4 kg, which is similar to the 116 kg calculated by Wroe et al (1999) on the basis of femoral variables (but not 58 kg on the basis of the circumference of the femur; Wroe et al. 2004a). Using other postcranial equations Argot (2004c) obtained somewhat different values (47.5 kg–49.5 kg, 82 kg–86.7 kg, and 108 kg for the holotype). Larger values were obtained using condylobasal skull length (150 kg; Sorkin 2008), while highly unlikely lower estimates were predicted from endocranial volume (26 kg; Wroe et al. 2003) and tooth variables (30.2 kg; Prevosti et al. 2013).

The study of the postcranial skeleton indicated terrestrial progression with incipient cursoriality (Riggs 1934; Ercoli et al. 2012), possibly an ambush predator, attacking by surprise rather than the chase (Goin and Pascual 1987; Argot 2004a, c). Riggs (1934) and Argot (2004c) concluded that the forelimbs were digitigrade or semidigitigrade, while the hindlimbs were plantigrade. The forelimbs would have had manipulative capabilities to capture and secure prey (Argot 2004a, b). The neck was longer than in other sparassodonts, more flexible and strongly muscled (Argot 2004a, c).

Thylacosmilus atrox was hypercarnivorous but with a bite force extremely low compared to other sparassodonts (Wroe et al. 2004b; Blanco et al. 2011). Geometric morphometric studies suggested that the cranium shared a morphospace similar to *Barbourofelis* (Prevosti et al. 2010), another eutherian sabertooth.

Several functional studies have discussed the predation behaviour of *T. atrox* compared to eutherian saber-toothed cats. Most studies have adopted the stabbing model, in which the primarily force applied to the canines was neck driven (e.g., Marshall 1976c; Turnbull 1978; Churcher 1985; Argot 2004a, c; Wroe et al. 2004b, 2013), rather than the primary force coming from jaw adductors, as in saber-toothed cats (i.e., the canine-sharing bite model following Wroe et al. 2004b, 2013) (e.g., Goin and Pascual 1987; Therrien 2005). A recent 3D finite element analysis has demonstrated that the jaw adductors played an insignificant role in the killing bite (Wroe et al. 2013). A maximal gape of 105.8° was inferred for *T. atrox*, which is much larger than for any saber-toothed cat (Wroe et al. 2013). Goin and Pascual (1987) considered the high ratio between length and width of the upper canines and were in favor of long and shallow wounds on vital, bone-free body surfaces of the prey, such as abdomen and throat with canines fuctioning as guides during molar occlusion.

Comments. *Thylacosmilus atrox* had a morphology that departs from other sparassodonts. It had a short and massive skull with complete postorbital bar, and very small nasals as seen in dorsal view, partially covered by the maxilla. The large saber-like upper canine was ever-growing and deeply anchored in the maxilla

(Riggs 1934; Marshall 1976c; Turnbull and Segall 1984; Goin and Pascual 1987). There has been discussion of the dental formula. One specimen clearly possessed at least one pair of lower incisors—possibly more than one—(Goin and Pascual 1987). Wear facets on the lower incisors suggested that uppers were also present (Churcher 1985). *Thylacosmilus atrox* had two premolars in each jaw; the last upper premolar has been interpreted as a retained deciduous element (Goin and Pascual 1987; Forasiepi and Sánchez-Villagra 2014).

Goin and Pascual (1987) suggested that all known late Miocene–Pliocene thylacosmilid remains belong to a single species. Citing Article 23a of the International Code of Zoological Nomenclature, the authors suggested maintaining the broadly known name of the junior synonym *Thylacosmilus atrox* Riggs 1933, instead of *Achlysictis lelongi* Ameghino 1891.

Borhyaenidae Ameghino 1894

Borhyaenidae are large-sized sparassodonts with massive skulls and dentaries strongly attached or fused at symphysis, resembling the cranial architecture of the Tasmanian devil (*Sarcophilus harrisi*). All taxa may have had hypercarnivorous diet (Prevosti et al. 2013). Borhyaenidae includes the common ancestor of *Borhyaena, Arctodictis,* and all its descendants (e.g., Forasiepi 2009; Forasiepi et al. 2015). The oldest borhyaenid, *Australohyaena antiquua*, comes from the late Oligocene (Deseadan) while the youngest from the middle Miocene (Colloncuran) (Table 3.1). Classically, several Paleogene sparassodonts were included within the Borhyaenidae; however, we use here a more restricted definition. A putative *Borhyaena* sp. has been identified for the late Miocene (Huayquerian) (Marshall 1978). If correct, the stratigraphic range of the group should be extended another ∼10 Ma over the range indicated here.

Acrocyon Ameghino 1887
(Tables 3.1 and 3.2)

Included species. *Acrocyon sectorius* Ameghino 1887 (type species) and *A. riggsi* (Sinclair 1930).

Temporal and geographic distribution. *Acrocyon riggsi* and *A. sectorius* come from the early Miocene (Colhuehuapian and Santacrucian, respectively), Argentina.

Paleoecology. The two *Acrocyon* species overlapped in size. Body mass estimations for *A. sectorius* on the basis of tooth variables provided 28.7 kg and 16.26 kg (Wroe et al. 2004a; Prevosti et al. 2013, respectively), while *A. riggsi* resulted in 26.3 kg and 17 kg (Wroe et al. 2004a; Prevosti et al. 2013, respectively).

Comments. Oiso (1991) questionably referred one poorly preserved specimen from the middle Miocene (Colloncuran) of Nazareno (Bolivia) to *Acrocyon* sp. Alternatively, Croft et al. (in press) suggested that the specimen possibly belongs to a new species also present in Cerdas (Bolivia) in outcrops of comparable age.

Acrocyon species are very similar in size and morphology to the contemporaneous species of *Borhyaena*. The likely possibility that *Acrocyon* represents part of the intraspecific variability of *Borhyaena* should be explored (Forasiepi 2009).

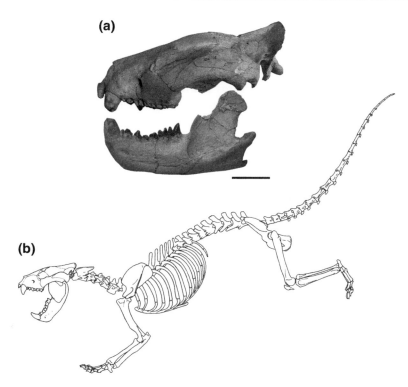

Fig. 3.12 *Arctodictis sinclairi* (MLP 85-VII-3-1), from the early Miocene (Colhuehuapian Age), Gran Barranca, Patagonia (Argentina), cranium and left dentary in lateral view (**a**); restoration of the skeleton (modified from Forasiepi 2009) (**b**). Scale bar = 5 cm

Arctodictis Mercerat 1891
(Figs. 3.7a and 3.12; Tables 3.1 and 3.2)

Included species. *Arctodictis munizi* Mercerat 1891 (type species) and *A. sinclairi* Marshall 1978.

Temporal and geographic distribution. *Arctodictis sinclairi* and *A. munizi* come from the early Miocene (Colhuehuapian and Santacrucian, respectively), Argentina.

Paleoecology. *Arctodictis munizi* was the largest post-Deseadan sparassodont with linear skull measurements similar to a lion. Based on the available material, the body mass of *A. munizi* seems to be underestimated (e.g., 51.6 kg in Wroe et al. 2004a; 37 kg in Vizcaíno et al. 2010; 43 kg in Prevosti et al. 2013, based on tooth equations). The older *A. sinclairi* is about 20% smaller (in linear cranial measurements) than the Santacrucian taxon, and its body mass was estimated at ∼40 kg using the centroid size of the humerus and ulna (Ercoli and Prevosti 2011). Estimations for *A. sinclairi* using tooth variables appear to be underestimated (e.g., 23.3 kg in Wroe et al. 2004a; 18.34 kg in Prevosti et al. 2013).

Arctodictis sinclairi was considered a generalized terrestrial sparassodont with plantigrade posture (Forasiepi 2009; Ercoli et al. 2012). Both *Arctodictis* species have dentitions that suggested a hypercarnivorous diet (Prevosti et al. 2013). In particular, *A. munizi* was considered a bone-cracker (Forasiepi et al. 2004).

Comments. *Arctodictis sinclairi* is known by an almost complete skeleton (Fig. 3.12; Forasiepi 2009). *Arctodictis* species have a similar skull, dentition, and postcranium to *Australohyaena antiquua* and *Borhyaena* species. Cladistic analyses suggested that those taxa shared a close common ancestor (Forasiepi et al. 2004, 2006, 2015; Forasiepi 2009; Engelman and Croft 2014; Suarez et al. 2015).

Australohyaena Forasiepi et al. 2015
(Tables 3.1 and 3.2)

Included species. *Australohyaena antiquua* (Ameghino 1894).
Temporal and geographic distribution. Late Oligocene (Deseadan), Argentina.
Paleoecology. *Australohyaena antiquua* was large (~ 70 kg of body mass) and robust (Forasiepi et al. 2015). The tooth morphology indicated a hypercarnivorous diet. The tooth equations together with the robustness of p3, vaulted skull, robustness of the jaws, and strong development of the temporal musculature (inferred from skull bony landmarks) suggested that *A. antiquua* was a bone-cracker and thus a hyena-like ecomorph (Forasiepi et al. 2015).

Comments. *Australohyaena antiquua* is represented by an almost complete skull and dentition. Phylogenetic reconstructions placed it close to the Miocene genus *Arctodictis* (Forasiepi et al. 2015).

Originally, the species was recognized as ?*Borhyaena antiqua* Ameghino 1894, then changed to *Proborhyaena antiqua* by Ameghino (1897), *Pharsophorus? antiquus* by Marshall (1978) and *Australohyaena antiqua* by Forasiepi et al. (2015). However, the spelling in Latin fem. sing. adj. is "antiquua." Under the provisions of ICZN art. 33.2, the name was suggested to correct *Australohyaena antiquua* (Babot and Forasiepi 2016).

Borhyaena Ameghino 1887
(Figs. 3.7b, c and 3.13; Tables 3.1 and 3.2)

Included species. *Borhyaena tuberata* Ameghino 1887 (type species) and *B. macrodonta* (Ameghino 1902b).
Temporal and geographic distribution. *Borhyaena macrodonta* comes from the early Miocene (Colhuehuapian), Argentina; *B. tuberata* from the early to middle Miocene (Santacrucian and Friasian), Argentina and Chile.
Paleoecology. The body mass of *B. tuberata* was estimated on 36.4 kg on the basis of the centroid size of the ulna and tibia (Ercoli and Prevosti 2011). Slightly lower results were obtained by using equations based on the postcranium (e.g., ~ 23 kg in Argot 2003b; 21.4 kg in Wroe et al. 2004a) and the dentition (e.g., 23.31 kg or 28.5 kg, in Vizcaíno et al. 2010 and Prevosti et al. 2013, respectively). The body mass of *B. macrodonta* was estimated at 34.7 kg or 31.25 kg (Wroe et al. 2004a and Prevosti et al. 2013, respectively) using dental variables.

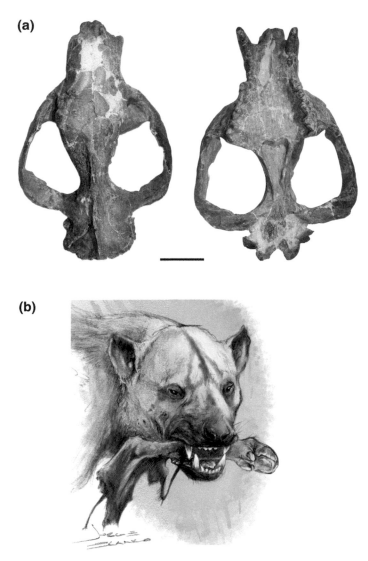

Fig. 3.13 *Borhyaena tuberata* (FMNH P 13252), from the early Miocene (Santacrucian Age), Patagonia (Argentina), cranium in dorsal and ventral views (**a**); artistic reconstruction by Jorge Blanco (**b**). Scale bar = 5 cm

The study of the postcranium of *B. tuberata* indicated that it was a terrestrial predator (Sinclair 1906; Marshall 1978; Argot 2003b, 2004a; Forasiepi 2009) with some cursorial capabilities (Argot 2003b; Ercoli et al. 2012). The limb architecture indicated parasagittal and more restricted pronation/supination movements, and semi- or fully digitigrade posture in the forelimb (Argot 2003b, 2004a).

Borhyaena species were hypercarnivorous (Prevosti et al. 2013). Analysis of mandibular force profiles, studied in *B. tuberata,* indicated a powerful canine bite with unpredictable direction of forces and bone breaking capabilities at the level of the last molar (Blanco et al. 2011).

Comments. *Borhyaena tuberata* is an iconic sparassodont. It is the first sparassodont described (Ameghino 1887) and the archetype for the group. *Borhyaena tuberata* is represented by several specimens including cranial and postcranial remains (Sinclair 1906).

A fragmentary dentary with the m2 (MACN-PV 13207) has been collected from the late Miocene (Huayquerian) beds of Entre Ríos (Argentina), which closely resembles *Borhyaena* (Marshall 1978). If the assignation to the genus is proved to be correct, the stratigraphic range of the taxon would notably increase.

Fredszalaya Shockey and Anaya 2008
(Tables 3.1 and 3.2)

Included species. *Fredszalaya hunteri* Shockey and Anaya 2008.
Temporal and geographic distribution. Late Oligocene (Deseadan), Bolivia.
Paleoecology. Large-sized taxon, possible hypercarnivorous.

Comments. *Fredszalaya hunteri* is a Neotropical sparassodont, hitherto found only in the late Oligocene (Deseadan) of Bolivia. Shockey and Anaya (2008) suggested affinities between *F. hunteri* and *Borhyaena.* This hypothesis awaits testing with cladistic analysis.

3.2.1 Problematic Taxa

Some sparassodont species are based on scarce and fragmentary material, and consequently the diagnoses are indeterminate. In other cases, the holotypes are missing from collections. As a result, the following taxa are regarded as nomina dubia.

Angelocabrerus Simpson 1970
(Tables 3.1 and 3.2)

Included species. *Angelocabrerus daptes* Simpson 1970 (nomen dubium).
Temporal and geographic distribution. Middle Eocene (Casamayoran), Argentina.

Comments. *Angelocabrerus daptes* is based on a single, much worn specimen in the MMP collections, which was described and illustrated by Simpson (1970: Figs. 1–3). Because of the condition of the specimen, no diagnostic features can be recognized. The specimen can no longer be located (Dondas 2015 com. pers.). *Angelocabrerus daptes* was about the size of *Borhyaena* and *Arminiheringia* (~20.5 kg; Zimicz 2012). According to Simpson, this taxon is closer in morphology to the geologically younger *Borhyaena* than the contemporaneous *Arminiheringia* (Simpson 1970). It was likely hypercarnivorous (Zimicz 2012). Because the holotype is lost, we consider this taxon as nomen dubium.

Argyrolestes Ameghino 1902a
(Tables 3.1 and 3.2)

Included species. *Argyrolestes peralestinus* Ameghino 1902a (nomen dubium).
Temporal and geographic distribution. Middle Eocene (Casamayoran),
Argentina.

Comments. The species is known only by its holotype, a broken upper molar
(Ameghino 1902a; Simpson 1948; Marshall 1978) collected from Patagonia.
Currently, the material is lost from the MACN collections (Alvarez 2013, com.
pers.). According to Marshall (1978), the specimen might correspond to one of the
upper molars in the dentition of the contemporaneous *Nemolestes spalacotherinus*.
Because of these uncertainties and loss of the holotype, we consider the taxon as
nomen dubium.

Eutemnodus Burmeister 1885
(Tables 3.1 and 3.2)

Included species. *Eutemnodus americanus* Bravard 1858 (nomen dubium).
Temporal and geographic distribution. Late Miocene (Huayquerian), Argentina.
Comments. Only one species, *Eutemnodus americanus*, based on isolated upper
molars is currently recognized (Forasiepi et al. 2007). In the revision of the borhyaenids, Marshall (1978) tentatively recognized the species *E. acutidens* and *E.
propampinus*. In a later revision, the specimens included in these species were
identified in different taxonomic groups (Forasiepi et al. 2007). The diagnosis of
Eutemnodus is almost indeterminate (Marshall 1978), and for this reason we consider the taxon as nomen dubium. *Eutemnodus americanus* is interesting in that, if
valid, it records the presence of a borhyaenid-like sparassodont in the late Miocene
(Marshall 1978; Forasiepi et al. 2007).

3.2.2 Tiupampan Taxa

The finding of metatherians from the early Paleocene in Tiupampa, Bolivia, including
exquisite cranial and postcranial material, provided new data on the early radiation
of the group in South America. *Mayulestes ferox* and *Allqokirus australis* were
claimed to be the earliest sparassodonts (e.g., de Muizon 1994, 1998; de Muizon
et al. 1997; de Muizon 1998), or these two taxa in combination with *Pucadelphys
andinus*, *Andinodelphys cochabambensis* and *Jaskhadelphys minutus* as recently
suggested by Muizon et al. (2015). This hypothesis requires support from cladistic
analyses. The "key" synapomorphy used to define the Sparassodonta (including
Mayulestes) was the presence of a medial process of the squamosal projecting
medially, nearly reaching the foramen ovale (de Muizon et al. 1997; de Muizon
1998). This process is clearly seen in *Mayulestes* (de Muizon 1998: Fig. 8), but its
presence and universality among sparassodonts is ambiguous (Forasiepi 2009).

According to Goin (2003), the molar structure of *Mayulestes* is primitive and reminiscent of the Peradectoidea.

Phylogenetic hypotheses have recovered the Tiupampan taxa (i.e., *Mayulestes, Pucadelphys,* and *Andinodelphys*) as a monophyletic group, placed among stem Marsupialia (e.g., Rougier et al. 1998; Babot 2005; Forasiepi 2009 and derived matrixes: Engelman and Croft 2014; Forasiepi et al 2015; Suarez et al. in press). This hypothesis (Fig. 3.2) implies another parallel corollary: The South American metatherians are represented by lineages of diverse origin that invaded South America more than once and whose ancestral forms were Holarctic (Forasiepi 2009; see also Case et al. 2005; Goin et al. 2016). In turn, the crown group Marsupialia diverged in South America and represents a relict of one of these metatherian lineages that radiated in the continent, later invading Antarctica and Australia. By contrast, Metatheria is Holarctic in origin, with North America (Simpson 1950; Lillegraven 1969; Patterson and Pascual 1968, 1972; Tedford 1974; Keast 1977) or Asia (Luo et al. 2003) the probable area where the basal cladogenesis of the group took place (Fig. 3.2).

In summary, currently the accepted earliest sparassodonts are recorded in the early Eocene (Itaboraian), with a radiation of the group during the middle Eocene (Casamayoran) (Babot 2005; Forasiepi 2009; Babot and Forasiepi 2016). A large revision would required if the putative sparassodonts from Tiupampa were certainly members of Sparassodonta: The stratigraphic time span involved more than 10 Ma as currently understood, beginning with the start of the Cenozoic, and a much primitive metatherian morphology would have to define the group.

References

Ameghino F (1887) Enumeración sistemática de las especies de mamíferos fósiles coleccionados por Carlos Ameghino en los terrenos eocenos de la Patagonia Austral y depositados en el Museo de La Plata. Bol Mus La Plata 1:1–26

Ameghino F (1889) Contribución al conocimiento de los mamíferos de fósiles de la República Argentina, obra escrita bajo los auspicios de la Academia Nacional de Ciencias de la República Argentina para presentarla Exposición Universal de París de 1889. Acad Nac Cien Córdoba Actas 6:1–1027

Ameghino F (1891) Nuevos restos de mamíferos fósiles descubiertos por Carlos Ameghino en el Eoceno Inferior de la Patagonia austral. Especies nuevas, adiciones y correcciones. Rev Arg Hist Nat 1:289–328

Ameghino F (1894) Enumération synoptique des espèces de mammifères fossiles des formations éocènes de Patagonie. Acad Nac Cien Córdoba Bol 13:259–452

Ameghino F (1897) Mammifères Crétacés de l'Argentine. Deuxième contribution à la connaissance de la faune mammalogique des conches à *Pyrotherium*. Bol Inst Geogr Arg 18:406–521

Ameghino F (1902a) Notice preliminaires sur les mammifères nouveaux des terraines Crétacés de Patagonie. Acad Nac Cien Córdoba Bol 17:5–70

Ameghino F (1902b) Première contribution à la connaissance de la faune Mammalogique des couches à *Colpodon*. Acad Nac Cien Córdoba Bol 17:71–141

Ameghino F (1904) Nuevas especies de mamíferos Cretáceos y Terciarios de la República
    Argentina. Anales Soc Ci Argent 56:193–208; 57:162–175, 327–341; 58:35–41, 57–71, 182–
    192, 225–291
Argot C (2003a) Postcranial functional adaptations in the South American Miocene borhyaenoids
    (Mammalia, Metatheria): *Cladosictis, Pseudonotictis*, and *Sipalocyon*. Alcheringa 27:303–356
Argot C (2003b) Functional adaptations of the postcranial skeleton of two Miocene borhyaenoids
    (Mammalia, Metatheria) *Borhyaena* and *Prothylacynus*, from South America. Palaeontology
    46:1213–1267
Argot C (2004a) Evolution of South American mammalian predators (Borhyaenoidea): anatomical
    and palaeobiological implications. Zool J Linn Soc 140:487–521
Argot C (2004b) Functional-adaptative analysis of the postcranial skeleton of a Laventan
    borhyaenoid, *Lycopsis longirostris* (Marsupialia, Mammalia). J Vertebr Paleontol 24:689–708
Argot C (2004c) Functional-adaptive features and paleobiologic implications of the postcranial
    skeleton of the late Miocene sabretooth borhyaenoid, *Thylacosmilus atrox* (Metatheria).
    Alcheringa 28:229–266
Argot C, Babot MJ (2011) Postcranial morphology, functional adaptations and palaeobiology of
    Callistoe vincei, a predaceous metatherian from the Eocene of Salta, north-western Argentina.
    Palaeontology 54:447–480
Babot MJ (2005) Los Borhyaenoidea (Mammalia, Metatheria) del Terciario inferior del noroeste
    argentino. Aspectos filogenéticos, paleobiolólogicos y bioestratigráficos. Unpublished PhD.
    thesis, Universidad Nacional de Tucumán
Babot J, Forasiepi AM (2016) Mamíferos predadores nativos del Cenozoico sudamericano:
    evidencias filogenéticas y paleoecológicas. In: Agnolin FA, Lio GL, Brissón Egli F,
    Chimento NR, Novas FE (eds) Historia Evolutiva y Paleobiogeográfica de los Vertebrados de
    América del Sur.Contribuciones MACN, vol 6. Museo Argentino de Ciencias Naturales
    "Bernardino Rivadavia", Buenos Aires, pp 219–230
Babot MJ, Ortiz PE (2008) Primer registro de Borhyaenoidea (Mammalia, Metatheria,
    Sparassodonta) en la provincia de Tucumán (Formación India Muerta, Grupo Choromoro;
    Mioceno tardío). Acta Geol Lillo 21:34–48
Babot MJ, Powell JE, de Muizon C (2002) *Callistoe vincei*, a new Proborhyaenidae
    (Borhyaenoidea, Metatheria, Mammalia) from the early Eocene of Argentina. Geobios
    35:615–629
Bergqvist LP, Lima Moreira A, Ribeiro Pinto D (2006) Bacia de São José de Itaboraí. 75 Anos de
    História e Ciência. Serviço Geológico do Brasil, Rio de Janeiro
Blanco E, Jones WW, Grinspan GA (2011) Fossil marsupial predators of South America
    (Marsupialia, Borhyaenoidea): bite mechanics and palaeobiological implications. Alcheringa
    31:377–387
Bond M, Pascual R (1983) Nuevos y elocuentes restos craneanos de *Proborhyaena gigantea*
    Ameghino 1897 (Marsupialia, Borhyanidae, Proborhyaeninae) de la edad Deseadense. Un
    ejemplo de coevolución. Ameghiniana 20:47–60
Bond M, Tejedor MF, Campbell KE, Chornogubsky L, Novo N, Goin FJ (2015) Eocene primates
    of South America and the African origins of New World monkeys. Nature 520:538–541
Bravard A (1858) Monografía de los Terrenos Terciarios de las Cercanías del Paraná. Imprenta del
    Registro Oficial, Paraná. Reimpresión 1995, Imprenta del Congreso de la Nación, Buenos
    Aires
Burmeister G (1885) Examen crítico de los mamíferos y reptiles fósiles denominados por D.
    Augusto Bravard y mencionados en su obra precedente. Anales Mus Nac Hist Nat Buenos
    Aires 3:95–174
Cabrera A (1927) Datos para el conocimiento de los dasyuroideos fósiles argentinos. Rev Mus La
    Plata 30:271–315
Case JA, Goin FJ, Woodburne MO (2005) "South American" marsupials from the late Cretaceous
    of North America and the origin of marsupial cohorts. J Mammal Evol 12:461–494
Churcher CS (1985) Dental functional morphology in the marsupial sabre-tooth *Thylacosmilus
    atrox* (Thylacosmilidae) compared to that of felid sabre-tooths. Aust Mammal 8:201–220

Contreras VH (1990) Un Nuevo Hathlyacyninae (Mammalia: Borhyaenidae) del Chasiquense (Mioceno Superior) de la provincial de San Juan, Argentina. V Congreso Argentino de Paleontología y Bioestratigrafía. Actas, Tucumán, pp 163–168

Croft DA (2001) Cenozoic environmental change in South America as indicated by mammalian body size distributions (cenograms). Divers Distrib 7:271–287

Croft DA (2006) Do marsupials make good predators? Insights from predator–prey diversity ratios. Evol Ecol Res 8:1192–1214

Croft DA, Carlini AA, Ciancio MR, Brandoni D, Drew NE, Engelman RK, Anaya F (in press) New mammal faunal data from Cerdas, Bolivia, a middle-latitude Neotropical site that chronicles the end of the middle Miocene Climatic Optimum in South America. J Vertebr Paleontol

de Muizon C (1994) A new carnivorous marsupial from the Paleocene of Bolivia and the problem of the marsupial monophyly. Nature 370:208–211

de Muizon C (1998) *Mayulestes ferox*, a borhyaenoid (Metatheria, Mammalia) from the early Palaeocene of Bolivia. Phylogenetic and palaeobiologic implications. Geodiversitas 20:19–142

de Muizon C (1999) Marsupial skulls from the Deseadan (late Oligocene) of Bolivia and phylogenetic analysis of the Borhyaenoidea (Marsupialia, Mammalia). Geobios 32:483–509

de Muizon C, Argot C (2003) Comparative anatomy of the Tiupampa didelphimorphs; an approach to locomotory habits of early marsupials. In: Jones M, Dickman C, Archer M (eds) Predators with pouches: the biology of carnivorous marsupials. CSIRO Publishing, Victoria, pp 43–62

de Muizon C, Cifelli RL, Céspedes Paz R (1997) The origin of the dog-like borhyaenoid marsupials of South America. Nature 389:486–489

de Muizon C, Billet G, Argot C, Ladevèze S, Goussard F (2015) *Alcidedorbignya inopinata*, a basal pantodont (Placentalia, Mammalia) from the early Palaeocene of Bolivia: anatomy, phylogeny and palaeobiology. Geodiversitas 37:397–634

de Paula Couto C (1952) Fossil mammals from the beginning of the Cenozoic in Brazil. Marsupialia: Polydolopidae and Borhyaenidae. Am Mus Novit 1559:1–27

Degrange FJ, Tambussi CP, Moreno K, Witmer LW, Wroe S (2010) Mechanical analysis of feeding behavior in the extinct "terror bird" *Andalgalornis steulleti* (Gruiformes: Phorusrhacidae). PLoS One 5:e1 1856

Echarri S, Prevosti FJ (2015) Differences in mandibular disparity between extant and extinct species of metatherian and placental carnivore clades. Lethaia 48:196–204

Engelman RK, Croft DA (2014) A new species of small-bodied sparassodont (Mammalia: Metatheria) from the middle Miocene locality of Quebrada Honda, Bolivia. J Vert Paleontol 34:672–688

Engelman RK, Anaya F, Croft DA (2015) New specimens of *Acyonmyctoderos* (Metatheria, Sparassodonta) from Quebrada Honda, Bolivia. Ameghiniana 52:204–225

Ercoli MD, Prevosti FJ (2011) Estimación de masa de las especies de Sparassodonta (Metatheria, Mammalia) de la Edad Santacrucense (Mioceno Temprano) a partir de tamaños de centroide de elementos apendiculares: inferencias paleoecológicas. Ameghiniana 48:462–479

Ercoli MD, Prevosti FJ, Álvarez A (2012) Form and function within a phylogenetic framework: locomotory habits of extant predators and some Miocene Sparassodonta (Metatheria). Zool J Linn Soc 165:224–251

Ercoli MD, Prevosti FJ, Forasiepi AM (2014) The structure of the mammalian predator guild in the Santa Cruz Formation (late early Miocene), Patagonia, Argentina. J Mammal Evol 21:369–381

Esteban G, Nasif N, Georgieff SM (2014) Cronobioestratigrafía del Mioceno tardío-Plioceno temprano, Puerta de Corral Quemado y Villavil, provincia de Catamarca, Argentina. Acta Geol Lillo 26:165–192

Forasiepi AM (2009) Osteology of *Arctodictis sinclairi* (Mammalia, Metatheria, Sparassodonta) and phylogeny of Cenozoic metatherian carnivores from South America. Monografías Mus Arg Sci Nat "Bernardino Rivadavia" [ns] 6:1–174

Forasiepi AM, Carlini AA (2010) New thylacosmilid (Mammalia, Metatheria, Sparassodonta) from the Miocene of Patagonia, Argentina. Zootaxa 2552:55–68

Forasiepi AM, Sánchez-Villagra MR (2014) Heterochrony, dental ontogenetic diversity and the circumvention of constraints in marsupial mammals and extinct relatives. Paleobiology 40:222–237

Forasiepi AM, Goin FJ, di Martino V (2003) Una nueva especie de *Lycopsis* (Metatheria, Prothylacyninae) de la Formación Arroyo Chasicó (Mioceno Tardío), de la Provincia de Buenos Aires. Ameghiniana 40:249–253

Forasiepi AM, Goin FJ, Tauber AA (2004) Las especies de *Arctodictis* Mercerat 1891 (Metatheria, Borhyaenidae), grandes carnívoros del Mioceno de América del Sur. Rev Esp Paleontol 19:1–22

Forasiepi AM, Sánchez-Villagra MR, Goin FJ, Takai M, Kay RF, Shigehara N (2006) A new Hathliacynidae (Metatheria, Sparassodonta) from the middle Miocene of Quebrada Honda, Bolivia. J Vert Paleontol 26:670–684

Forasiepi AM, Martinelli AG, Goin FJ (2007) Revisión taxonómica de Parahyaenodon argentinus Ameghino y sus implicancias en el conocimiento de los grandes mamíferos carnívoros del Mio-Plioceno de América del Sur. Ameghiniana 44:143–159

Forasiepi AM, Babot MJ, Zimicz N (2015) *Australohyaena antiqua* (Mammalia, Metatheria, Sparassodonta), a large predator from the late Oligocene of Patagonia, Argentina. J Syst Palaeontol 13:503–525

Goin FJ (1997) New clues for understanding Neogene marsupial radiations. In: Kay RF, Madden RH, Cifelli RL, Flynn JJ (eds) A history of the neotropical fauna. Vertebrate paleobiology of the Miocene in Colombia. Smithsonian Institution Press, Washington, pp 185–204

Goin FJ (2003) Early marsupial radiations in South America. In: Jones M, Dickman C, Archer M (eds) Predators with pouches. The biology of carnivorous marsupials. CSIRO Publishing, Victoria, pp 30–42

Goin FJ, Candela A (2004) New Paleogene marsupials from the Amazon Basin of Eastern Perú. In: Campbell KE Jr (ed) The Paleogene mammalian fauna of Santa Rosa, Amazonian Perú. Nat Hist Mus Los Angeles County, Science Series, vol 40, pp 15–60

Goin FJ, Pascual R (1987) News on the biology and taxonomy of the marsupials Thylacosmilidae (late Tertiary of Argentina). Anales de la Academia Nacional de Ciencias Exactas Físicas y Naturales de Buenos Aires 39:219–246

Goin FJ, Palma RM, Pascual R, Powell JE (1986) Persistencia de un primitivo Borhyaenidae (Mammalia, Marsupialia) en el Eoceno Temprano de Salta (Fm. Lumbrera, Argentina), aspectos geológicos y paleoambientales relacionados. Ameghiniana 23:47–56

Goin FJ, Abello MA, Bellosi E, Kay R, Madden R, Carlini AA (2007) Los Metatheria sudamericanos de comienzos del Neógeno (Mioceno Temprano, edad-mamífero Colhuehuapense). Parte I: Introducción, Didelphimorphia y Sparassodonta. Ameghiniana 44:29–71

Goin FJ, Abello MA, Chornogubsky L (2010) Middle tertiary marsupials from central Patagonia (early Oligocene of Gran Barranca): understanding South America's Grande Coupure. In: Madden RH, Carlini AA, Vucetich MG, Kay RF (eds) The paleontology of Gran Barranca: evolution and environmental change through the middle Cenozoic of Patagonia. Cambridge University Press, New York, pp 71–107

Goin FJ, Woodburne MO, Zimicz AN, Martin GM, Chornogubsky L (2016) A brief history of South American metatherians, evolutionary contexts and intercontinental dispersals. Springer, Dordrecht

Gordon CL (2003) A first look at estimating body size in dentally conservative marsupials. J Mammal Evol 10:1–21

Goswami A, Milne N, Wroe S (2011) Biting through constraints: cranial morphology, disparity, and convergence across living and fossil carnivorous mammals. Proc R Soc Lond [B] 278:1831–1839

Hoffstetter R, Petter G (1983) *Paraborhyaena boliviana* et *Andinogale sallensis,* deux marsupiaux (Borhyaenidae) nouveaux du Déséadien (Oligocène Inférieur) de Salla (Bolivie). C R Acad Sci 296:205–208

Keast A (1977) Historical biogeography of the marsupials. In: Stonehouse B, Gilnore D (eds) The biology of the marsupials. University Park Press, Baltimore, pp 69–95

Ladevèze S, de Muizon C (2007) The auditory region of Paleocene Pucadelphydae (Mammalia, Metatheria) from Tiupampa, Bolivia with phylogenetic implications. Palaeontology 50:1123–1154

Ladevèze S, de Muizon C (2010) Evidence of early evolution of Australidelphia (Mammalia, Metatheria) in South America: phylogenetic relationships of the metatherian from the late Paleocene of Itaboraí (Brazil) based on teeth and petrosal bones. Zool J Linn Soc 159:746–784

Lillegraven JA (1969) Latest Cretaceous mammals of upper part of Edmonton Formation of Alberta, Canada, and review of marsupial-placental dichotomy in mammalian evolution. Univ Kans Paleontol Contrib Pap 50:1–122

Linares OJ (2004) Bioestratigrafía de la fauna de mamíferos de las formaciones Socorro, Urumaco y Codore (Mioceno medio–Plioceno temprano) de la región de Urumaco, Falcón, Venezuela. Paleobiología Neotropical 1:1–23

Loomis FB (1914) The Deseado Formation of Patagonia. Amherst College, New Haven

Luo Z-X, Ji Q, Wible JR, Yuan CX (2003) An early Cretaceous tribosphenic mammal and metatherian evolution. Science 302:1934–1940

Marshall LG (1976a) New didelphine marsupials from the La Venta fauna (Miocene) of Colombia, South America. J Paleo 50:402–418

Marshall LG (1976b) A new borhyaenid (Marsupialia, Borhyaeninae) from the Arroyo Chasicó Formation (lower Pliocene), Buenos Aires Province, Argentina. Ameghiniana 13:289–299

Marshall LG (1976c) Evolution of the Thylacosmilidae, extinct saber-tooth marsupials of South America. PaleoBios 23:1–30

Marshall LG (1977a) Evolution of the carnivorous adaptative zone in South America. In: Hecht MK, Goody PC, Hecht BM (eds) Major patters in vertebrate evolution. Plenum Press, New York, pp 709–721

Marshall LG (1977b) A new species of *Lycopsis* (Borhyaenidae, Marsupialia) from the La Venta Fauna (Miocene) of Colombia, South America. J Paleontol 51:633–642

Marshall LG (1978) Evolution of the Borhyaenidae, extinct South American predaceous marsupials. Univ Calif Publ Geol Sci 117:1–89

Marshall LG (1979) Review of the Prothylacyninae, an extinct subfamily of South American "dog-like" marsupials. Fieldiana. Geol [ns] 3:1–50

Marshall LG (1981) Review of the Hathlyacyninae, an extinct subfamily of South American "dog-like" marsupials. Fieldiana. Geol [ns] 7:1–120

Marshall LG, de Muizon C (1988) The dawn of the age of mammals in South America. Natl Geogr Res 4:23–55

Marshall LG, Patterson B (1981) Geology and geochronology of the mammal-bearing tertiary of the Valle de Santa Maria and Rio Corral Quemado, Catamarca Province, Argentina. Fieldiana. Geol [ns] 9:1–80

Marshall LG, Case JA, Woodburne MO (1990) Phylogenetic relationships of the families of marsupials. Curr Mammal 2:433–502

Martin GM, Tejedor MF (2007) Nueva especie de *Pseudonotictis* (Metatheria, Sparassodonta, Hathliacynidae). Ameghiniana 44:747–750

Mercerat A (1891) Caracteres diagnósticos de algunas especies de Creodonta conservadas en el Museo de La Plata. Revista Mus. La Plata 2:51–56

Mercerat A (1917) Notas sobre algunos carnívoros fósiles y actuales de la América del Sud. H. Errado y Cia. Impresores, Buenos Aires

Millien V, Bovy H (2010) When teeth and bones disagree: body mass estimation of a giant extinct rodent. J Mammal 91:11–18

Mones A (2014) *Contrerascynus*, new name for *Simpsonia* Contreras, 1990 (Mammalia, Sparassodonta, Hathliacynidae), *non* Rochebrune, 1904 (Bivalvia, Unionidae), *non* Baker, 1911 (Gastropoda, Lymnaeidae). Rev Bras Paleontol 17:435–436

Mones A, Rinderknecht A (2004) Primer registro de Thylacosmilidae en el Uruguay (Mammalia: Marsupialia: Sparassodonta). Comunicaciones Paleontológicas del Museo Nacional de Historia Natural y Antropología 34:193–200

Mones A, Ubilla M (1978) La edad Deseadense (Oligoceno Inferior) de la Formación Fray Bentos y su contenido paleontológico, con especial referencia a la presencia de *Proborhyaena cf. gigantea* Ameghino (Marsupialia, Borhyaenidae) en el Uruguay -nota preliminar. Comunicaciones Paleontológicas MHNM 7:151–157

Myers TJ (2001) Marsupial body mass prediction. Austral J Zool 49:99–118

Oiso Y (1991) New land mammal locality of middle Miocene (Colloncuran) age from Nazareno, Southern Bolivia. In: Suarez-Soruco R (ed) Fósiles y facies de Bolivia—vol 1 Vertebrados. Revista Técnica de YPF 12: 653–672

Pascual R, Bocchino A (1963) Un nuevo Borhyaninae (Marsupialia) del Plioceno Medio de Hidalgo (La Pampa). Ameghiniana 3:97–107

Patterson B, Marshall LG (1978) The Deseadan, early Oligocene, Marsupialia of South America. Fieldiana. Geol [ns] 41:37–100

Patterson B, Pascual R (1968) Evolution of mammals in southern continents. V. Fossil mammal fauna of South America. Quart Rev Biol 43:409–451

Patterson B, Pascual R (1972) The fossil mammal fauna of South America. In: Keast A, Erk FC, Glass B (eds) Evolution of mammals and Southern continents. State Univesity of New York Press, Albany

Petter G, Hoffstetter R (1983) Les marsupiaux du Déséadien (Oligocène Inférieur) de Salla (Bolivie). Annls Paléont (Vert–Invert) 69:175–234

Powell JE, Babot MJ, García-López DA, Deraco MV, Herrera CM (2011) Eocene vertebrates of northwestern Argentina: annotated list. In: Salfity JA, Marquillas RA (eds) Cenozoic geology of the Central Andes of Argentina. SCS Publisher, Salta, pp 349–370

Prevosti FJ, Turazzini GF, Chemisquy MA (2010) Morfología craneana en tigres dientes de sable: alometría, función y filogenia. Ameghiniana 47:239–256

Prevosti FJ, Turazzini GF, Ercoli MD, Hingst-Zaher E (2012a) Mandible shape in marsupial and placental carnivorous mammals: a morphological comparative study using geometric morphometrics. Zool J Linn Soc 164:836–855

Prevosti FJ, Forasiepi AM, Ercoli MD, Turazzini GF (2012b) Paleoecology of the mammalian carnivores (Metatheria, Sparassodonta) of the Santa Cruz Formation (late early Miocene). In: Vizcaíno SF, Kay RF, Bargo MS (eds) early Miocene paleobiology in Patagonia. Cambridge University Press, Cambridge, pp 173–193

Prevosti FJ, Forasiepi AM, Zimicz N (2013) The evolution of the Cenozoic terrestrial mammalian predator guild in South America: competition or replacement? J Mammal Evol 20:3–21

Reguero M, Candela AM (2011) Late Cenozoic mammals from Northwest of Argentina. In: Salfity JA, Marquillas RA (eds) Cenozoic geology of Central Andes of Argentina. Instituto del Cenozoico, Salta, pp 411–426

Reig OA (1957) Nota previa sobre los mamíferos de la Formación Chasicó. Ameghiniana 1:27–31

Riggs ES (1933) Preliminary description of a new marsupial saber-tooth from the Pliocene of Argentina. Field Mus Nat Hist Geol ser 6:61–66

Riggs ES (1934) A new marsupial saber-tooth from the Pliocene of Argentina and its relationships to other South American predaceous marsupials. T Am Philos Soc [ns] 24:1–31

Roth S (1903) Noticias sobre nuevos mamíferos fósiles del Cretáceo superior y Terciario de la Patagonia. Revista Mus. La Plata 11:133–158

Rougier GW, Wible JR, Novacek MJ (1998) Implications of *Deltatheridium* specimens for early marsupial history. Nature 396:459–463

Rougier GW, Wible JR, Novacek MJ (2004) New specimen of *Deltatheroides cretacicus* (Metatheria, Deltatheroidea) from the late Cretaceous of Mongolia. Bull Carnegie Mus Nat Hist 36:245–266

Sánchez-Villagra MR (2013) Why are there fewer marsupials than placentals? On the relevance of geography and physiology to evolutionary patterns of mammalian diversity and disparity. J Mammal Evol 20:279–290

Scheyer TM, Aguilera OA, Delfino M, Fortier DC, Carlini AA, Sánchez R et al (2013) Crocodylian diversity peak and extinction in the late Cenozoic of the northern Neotropics. Nat. Commun 4:1907. doi:10.1038/ncomms2940

Scott WB (1937) A history of land mammals in the Western Hemisphere (revised edition). The MacMillan Company, New York

Sears K (2004) Constraints on the morphological evolution of marsupial shoulder girdles. Evolution 58:2353–2370

Shockey BJ, Anaya F (2008) Postcranial osteology of mammals from Salla, Bolivia (late Oligocene): form, function, and phylogenetic implications. In: Sargis EJ, Dagosto M (eds) Mammalian evolutionary morphology. A tribute to Frederick S. Szalayc. Springer, Netherlands, pp 135–157

Simpson GG (1930) Post-Mesozoic marsupialia. In: Junk W (ed) Fossilium catalogus. I: Animalia. W. Junk, Berlin, pp 1–87

Simpson GG (1935) Descriptions of the oldest known South American mammals from the Río Chico Formation. Amer Mus Novit 793:1–25

Simpson GG (1948) The beginning of the age of mammals in South America, Part 1. Bull Amer Mus Nat Hist 91:1–232

Simpson GG (1950) History of the fauna of Latin America. Am Sci 38:361–389

Simpson GG (1970) Mammals from the early Cenozoic of Chubut, Argentina. Breviora 360:1–13

Simpson GG (1971) The evolution of marsupials in South America. An Acad Bras Cien 43:103–118

Simpson GG (1980) Splendid isolation: the curious history of South American mammals. Yale Univesity Press, New Haven

Sinclair WJ (1906) Mammalia of the Santa Cruz beds: Marsupialia. Rep Princeton Univ Exped Patagonia, 1896–1899, 4 (Paleontology). Princeton

Sinclair WJ (1930) New carnivorous marsupials from the Deseado Formation of Patagonia. Field Mus Nat Hist Geol Ser 1:35–39

Sorkin B (2008) A biomechanical constraint on body mass in terrestrial mammalian predators. Lethaia 41:333–347

Suarez C, Forasiepi AM, Goin FJ, Jaramillo C (2015) Insights into the Neotropics prior to the Great American Biotic Interchange: new evidence of mammalian predators from the Miocene of Northern Colombia. J Vert Paleontol. doi:10.1080/02724634.2015.1029581

Szalay FS (1994) Evolutionary history of the marsupials and an analysis of osteological characters. Cambridge University Press, New York

Tedford RH (1974) Marsupials and the new paleogeography. In: Ross CA (ed) Paleogeographic provinces and provinciality. Society of Economic Paleontologists and Mineralogists, Special Publication 21:109–126

Therrien F (2005) Feeding behaviour and bite force of sabretoothed predators. Zool J Linn Soc 145:393–426

Turnbull WD (1978) Another look at dental specialization in the extinct saber-toothed marsupial, *Thylacosmilus*, compared with its placental counterparts. In: Butler PM, Joysey KA (eds) Development, function and evolution of teeth. Academic Press, London, pp 399–414

Turnbull WD, Segall W (1984) The ear region of the marsupial sabertooth, *Thylacosmilus*: influence of the sabertooth lifestyle upon it, and convergence with placental sabertooths. J Morphol 181:239–270

Van Valkenburgh B (1991) Iterative evolution of hypercarnivory in canids (Mammalia: Carnivora): evolutionary interactions among sympatric predators. Paleobiology 17:340–362

Villarroel C, Marshall LG (1982) Geology of the Deseadan (early Oligocene) age estratos Salla in the Salla-Luribay Basin, Bolivia with description of new Marsupialia. Geobios Mem Spec 6:201–211

Villarroel C, Marshall LG (1983) Two new late Tertiary marsupials (Hathlyacyninae and Sparassocyninae) from the Bolivian Altiplano. J Paleo 57:1061–1066

Vizcaíno SF, Bargo MS, Kay RF, Fariña RA, GiacomoM Di, Perry JMG, Prevosti FJ, Toledo N, Cassini GH, Fernicola JC (2010) A baseline paleoecological study for the Santa Cruz

Formation (late–early Miocene) at the Atlantic coast of Patagonia, Argentina. Palaeogeogr Palaeoclimatol Palaeoecol 292:507–519

Wroe S, Myers TJ, Wells RT, Gillespie A (1999) Estimating the weight of the Pleistocene marsupial lion, *Thylacoleo carnifex* (Thylacoleonidae: Marsupialia): implications for the ecomorphology of a marsupial super-predator and hypotheses of impoverishment of Australian marsupial carnivore faunas. Aust J Zool 47:489–498

Wroe S, Myers TJ, Seebacher F, Kear B, Gillespie A, Crowther Salisbury S (2003) An alternative method for predicting body mass: the case of the Pleistocene marsupial lion. Paleobiology 29:403–411

Wroe S, Argot C, Dickman C (2004a) On the rarity of big fierce carnivores and primacy of isolation and area: tracking large mammalian carnivore diversity on two isolated continents. Proc R Soc Lond [B] 271:1203–1211

Wroe S, McHenry C, Thomason C (2004b) Bite club: comparative bite force in big biting mammals and the prediction of predatory behaviour in fossil taxa. Proc R Soc Lond [B] 272:619–625

Wroe S, Chamoli U, Parr WCH, Clausen P, Ridgely R et al (2013) Comparative biomechanical modeling of metatherian and placental saber-tooths: A different kind of bite for an extreme pouched predator. PLoS ONE 8(6):e66888

Zimicz AN (2012) Ecomorfología de los marsupiales paleógenos de América del Sur. Unpublished PhD. thesis, Universidad Nacional La Plata

Zimicz AN (2014) Avoiding competition: the ecological history of late Cenozoic metatherian carnivores in South America. J Mammal Evol 21:383–393

Chapter 4
South American Fossil Carnivorans (Order Carnivora)

Abstract Carnivora is a clade of mammalian predators that evolved in northern continents during the Paleocene, and since the Miocene have invaded the southern continents (i.e., Africa and South America). They evolved a large diversity and disparity of body forms and size, which allowed the occupation of many ecological niches. Carnivorans arrived in South America in the late Miocene, when Central America provided a land bridge, or an island chain that facilitated migration of initial mammalian groups including carnivorans. The first carnivorans in South America were procyonids, followed by mustelids and canids in the late Pliocene, and felids, mephitids, and ursids in the Pleistocene. Their high diversity and morphological disparity can be explained through a combination of repeated immigrations and radiations into empty ecological zones. Here we present a synthesis of the systematics, distribution, and paleocology of fossil terrestrial carnivorans of South America.

Keywords Canidae · Felidae · Mustelidae · Procyonidae · Ursidae Mephitidae · Immigration · Speciation

4.1 Introduction

Like domesticated animals, carnivorans are a very familiar group that has a long history of interactions with hominids during the Pleistocene, where they were both the hunters and the hunted, as depicted by Paleolithic carnivoran representations in cave paintings and sculptures (e.g., Kurtén 1968; Brain 1981; Clutton Brock 1996a; Morey 2010; Figs. 1.2–3). Their place in human life as both prey animals and competitors gave them a symbolic value that explains their inclusion in the cultures, religions, and fables of different times (Clutton Brock 1996a). Their symbolic importance, value as material resources, and the use of domestic carnivorans for different purposes (e.g., dog and cat; Clutton Brock 1996a; Morey 2010) explains the intense interactions between humans and this mammalian clade during the Quaternary. Different kinds of relationships between carnivorans and

© Springer International Publishing AG 2018 85
F.J. Prevosti and A.M. Forasiepi, *Evolution of South American Mammalian Predators During the Cenozoic: Paleobiogeographic and Paleoenvironmental Contingencies*, Springer Geology, https://doi.org/10.1007/978-3-319-03701-1_4

humans are included in art (e.g., cave paintings, pottery; Cardich 1979; Paunero et al. 2005; Gordillo 2010), myths, use of their remains in burials (Prates et al. 2010; Politis et al. 2014), hunting, and taming wild species, like the domestic dog (Schwartz 1997; Prates et al. 2010; Stahl 2013).

Carnivorans are a monophyletic group, and although they are primarily predators that live by hunting and eating other animals, a habit that is thought to have been present in their most recent common ancestor, some descendants became adapted to omnivorous or herbivorous diets (Hunt 1996; Flynn and Wesley-Hunt 2005; Flynn et al. 2010). The key synapomorphy of carnivorans is the presence of a pair of modified teeth (carnassials: upper fourth premolar and lower first molar), with long crests that function as scissors and are optimized for slicing meat (Ewer 1973; Van Valkenburgh 1989) (Fig. 4.1).

Carnivora is the crown group that contains the closest common ancestor of living carnivorans and all its descendants and consists of two large clades of living and fossil taxa. Feliformia includes felids (Felidae), Asiatic linsangs (Prionodontidae), hyaenids (Hyaenidae), "false" sabertooth cats (extinct Nimravidae and Barbourofelidae), palm civets (Nandiniidae), civets (Viverridae), mongooses (Herpestidae), and falanoucs (Eupleridae); Caniformia includes dogs (Canidae), bears (Ursidae), skunks (Mephitidae), weasels, otters, and relatives (Mustelidae), red pandas (Ailuridae), raccoons (Procyonidae), seals (Phocidae), sea lions (Otariidae), walruses (Odobenidae), and bear-dogs (extinct Amphicyonidae) (Hunt 1996; Morlo et al. 2004; Flynn et al. 2010; Eizirik et al. 2010; Nyakatura and Bininda-Emonds 2012). Seals, sea lions, and walruses form a clade that is called Pinnipedia that includes the carnivorans living in the sea. More basal carnivoran groups are the Viverravidae and the paraphyletic Miacidae, which are successive sister taxa of the monophyletic group that includes Feliformia and Caniformia (Flynn et al. 2010; Fig. 1.5). Miacidae and Viverravidae were historically included within the Order Carnivora (e.g., Hunt 1996), but now are excluded from crown-group Carnivora and included in the clade Carnivoramorpha (Flynn and Wesley-Hunt 2005; Flynn et al. 2010).

Recent phylogenetic studies show that carnivorans form a clade with pangolins and Creodonta (an extinct clade of predators), as their successive sister taxa and all are members of Laurasiatheria (including ungulates, bats, and cetaceans) (O'Leary et al. 2013; Fig. 1.5). However, Creodonta is paraphyletic and is divided in Hyaenodonta and Oxyaenodonta, with Hyaenodonta being the closest sister taxon of Carnivora (Solé and Smith 2013; Flynn and Wesley-Hunt 2005; Fig. 1.5).

Carnivorans have a long fossil record that indicates an origin in northern continents. The oldest records concerned viverravids found in the early Paleocene of North America (NA). This group was also present in Asia in the late Paleocene. Most living families of Caniformia are first recorded in the Eocene, while the crown group Feliformia is only recorded since the late Oligocene (Hunt 1996; Flynn and Wesley-Hunt 2005). Carnivorans invaded a wide range of habitats in most continents and seas, and consequently they display great taxonomic diversity, morphological disparity, and size.

The fossil record of taxonomic diversity in SA could be explained by a combination of succesive immigrations from Central America and local radiations

Fig. 4.1 Map of South America showing the localities (black symbols), or states, provinces, departments and regions (white symbols) mentioned in the text, where fossil carnivorans have been recorded. 1 Orocual (Venezuela); 2 La Calera (Ecuador); 3 La Chimba (Ecuador); 4 Cotocallo (Ecuador); 5 Punin (Ecuador); Loma Alta (Ecuador); 7 La Carolina (Ecuador); 8 Talara (Perú); 9 Huánuco department (Perú); 10 Junín department (Perú); 11 Arequipa department (Perú); 12 Tirapata (Perú); 13 Tarija (Bolivia); 14 Piaui state (Brasil); 15 Bahia State (Brasil); 16 Minas Gerais state (Brasil); 17 Mato Grosso state (Brazil); 18 Rio Grande do Sul (Brasil); 19 Formosa province (Argentina); 20 Catamarca province (Argentina); 21 Mendoza province (Argentina); 22 Córdoba province (Argentina); 23 Entre Ríos province (Argentina); 24 Buenos Aires province; 25 Lujan (Argentina); 26 La Pampa province (Argentina); 27 Valdivia (Chile); 28 Tierra del Fuego Island (Chile and Argentina); 29 Beagle Channel (Chile and Argentina). Broken line: northern limit of the Patagonian Region; Point line: northern and western limits of the Pampean Region. Squares: Quaternary; triangles: Pliocene; stars: late Miocene

enabled by the availability of ecological zones, plus later radiation of several clades in SA (Prevosti and Soibelzon 2012). The immigration of carnivorans to SA was facilitated by tectonic changes occurring in Panama that established intermittent terrestrial connections with Central America during the late Miocene (see Chap. 2). The pattern of carnivoran immigration was incremental. Procyonids arrived in the late Miocene, followed by weasels and foxes in the late Pliocene, and felids, otters,

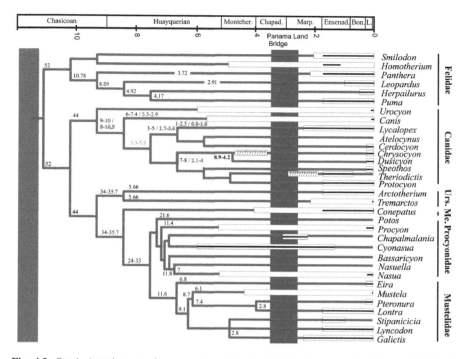

Fig. 4.2 South American carnivore supertree showing the fossil record of each genus in South America (black line) or other continents (white box). The numbers in the nodes correspond to molecular dates of divergence. For canids different estimates of the time of molecular divergence are given (tree nodes do not represent age of divergence). The orange box represents the "traditional" estimate of the establishment of the Panama Bridge, while the purple box indicates new evidence that suggest an early connection between South America and the Panama arc (see Chap. 2)

skunks, large canids, and bears in the Pleistocene (Fig. 4.2; Table 4.1; Soibelzon and Prevosti 2007, 2013; Prevosti and Soibelzon 2012). During the Late Pleistocene, other immigrations are recorded (e.g., *Urocyon* Baird 1857, *Canis dirus* Leidy 1858, *Smilodon fatalis* Leidy 1868), and it is possible that some carnivores originating in South America invaded Central America (e.g., *Speothos* Lund 1839, *Eira* Hamilton Smith 1842, *Procyon cancrivorus*). From the early–Middle Pleistocene (Ensenadan, 1.8–0.5 Ma) onwards, there was a large increase in diversity and morphological disparity, that was caused not only by immigration but also by the high rate of in situ speciation in the Pleistocene (Ensenadan–Lujanian; Prevosti and Soibelzon 2012). Molecular data suggest that the living populations of cougar (*Puma concolor* (Linnaeus 1771)) represent a recent (ca. 10 ka) recolonization of NA from SA (Culver et al. 2000). The fossil record shows a low level diversity, extinctions, speciation, and immigration during the late Miocene–Pliocene, with a substantial increase in these processes in the early–Middle Pleistocene (Ensenadan) and Late Pleistocene (Lujanian; Prevosti and Soibelzon 2012). In summary, most carnivoran groups immigrated to South

Table 4.1 Distribution of carnivorans in the South American Stages/Ages. ?: dubious record

	Huayquerian	Montehermosan	Chapadmalalan	Barrancalobian	Vorohuean	Sanandresian	Ensenadan	Bonaerian	Lujanian	Platan	Recent
"Canis" gezi							■				
"Felis" vorohuensis							■				
Arctotherium angustidens							■				
Arctotherium bonariense							?	■			
Arctotherium tarijense							?	■			
Arctotherium vetustum							?				
Arctotherium wingei							■				
Atelocynus microtis											■
Bassaricyon alleni											■
Bassaricyon bedarddi											■
Bassaricyon gabbi											■
Canis dirus									■		
Canis familiaris										■	■
Canis nehringi									■		
Cerdocyon thous									■	■	■
Chapalmalania altaefrontis			■								
Chapalmalania ortognatha				■							
Chrysocyon brachyurus							?	■	■	■	■
Conepatus chinga							?	■	■	■	■
Conepatus mercedensis							■				
Conepatus primaevus								?			
Conepatus semistriatus									■		■
Conepatus talarae									■		
Cyonasua argentina	■										
Cyonasua brevirostris	■										
Cyonasua clausa		■									
Cyonasua groeberi	?	?	?								
Cyonasua lutaria			■								
Cyonasua meranii							■				
Cyonasua pascuali	■										
Cyonasua sp. nov.	■										
Dusicyon australis											■
Dusicyon avus									■	■	

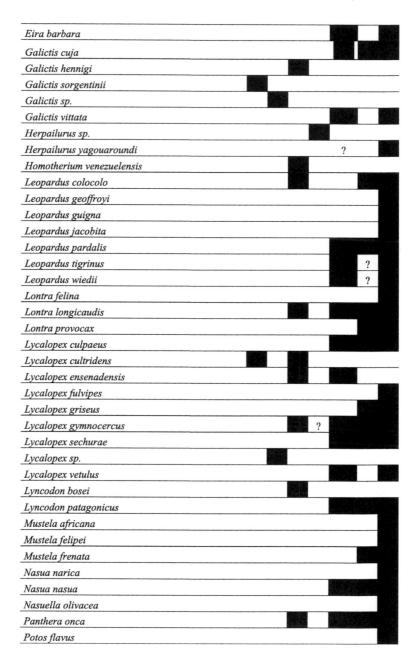

Eira barbara
Galictis cuja
Galictis hennigi
Galictis sorgentinii
Galictis sp.
Galictis vittata
Herpailurus sp.
Herpailurus yagouaroundi
Homotherium venezuelensis
Leopardus colocolo
Leopardus geoffroyi
Leopardus guigna
Leopardus jacobita
Leopardus pardalis
Leopardus tigrinus
Leopardus wiedii
Lontra felina
Lontra longicaudis
Lontra provocax
Lycalopex culpaeus
Lycalopex cultridens
Lycalopex ensenadensis
Lycalopex fulvipes
Lycalopex griseus
Lycalopex gymnocercus
Lycalopex sechurae
Lycalopex sp.
Lycalopex vetulus
Lyncodon bosei
Lyncodon patagonicus
Mustela africana
Mustela felipei
Mustela frenata
Nasua narica
Nasua nasua
Nasuella olivacea
Panthera onca
Potos flavus

Potos sp.		
Procyon cancrivorus		
Protocyon scagliorum		
Protocyon tarijensis	?	
Protocyon troglodytes		
Pteronura brasiliensis		
Puma concolor		
Smilodon fatalis		
Smilodon gracilis		
Smilodon populator	?	
Speothos pacivorus		
Speothos venaticus		
Stipanicicia pettorutii	?	
Theriodictis platensis		
Tremarctos ornatus		
Urocyon cinereoargenteus		
Urocyon sp.		

America, and speciated there, during the Pleistocene (Ensenadan-Platan). The fauna was further shaped by extinctions of large effect occurring at the end of the Ensenadan and Lujanian (Prevosti and Soibelzon 2012; see also Chap. 5).

4.2   Systematics, Distribution, and Paleoecology

Here we present a synthesis of the diversity, paleoecology, temporal and geographic distribution, and paleoecology of South American carnivores at the generic level (4.1). With regard to fossil occurrences of living taxa, we used known ecological aspects of living representatives to infer their paleoecology (Table 4.1). The large body of pertinent archeological works was not exhaustively searched, and some carnivoran records for the Holocene were probably missed. One issue with archeological papers is that they usually lack enough information to confirm the alleged taxonomic determinations.

Mammalia Linnaeus 1758
Carnivora Bowdich 1821
Feliformia Kretzoi 1945
Felidae Fischer 1817

Felids are highly specialized terrestrial apex predators with a hypercarnivorous diets (Ewer 1973; Wilson and Mittermeier 2009). They have a very short snout, powerful jaw muscles, and a reduced dentition that is specialized for shearing meat (Ewer 1973; Radinsky 1981; Biknevicius and Van Valkenburgh 1996). The oldest record of the family is dated between 35–28 Ma from Europe and spread to other

(a)

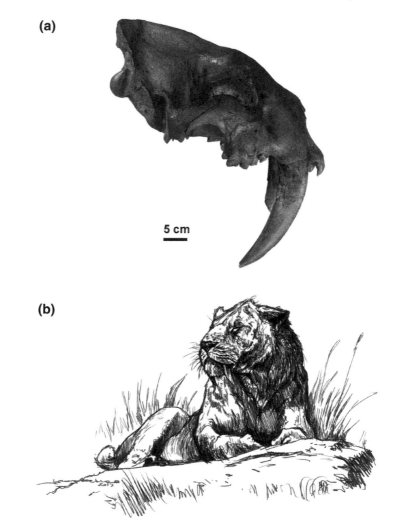

5 cm

(b)

Fig. 4.3 Lateral view of the skull of *Smilodon fatalis* (LACM-HC2001) (**a**) and life reconstruction of *Smilodon populator* (**b**). Scale = 5 cm

continents during the Miocene–Pleistocene (Hunt 1996; Werdelin et al. 2008). South American felids are included in the two subfamilies Machairodontinae and Felinae, which correspond to "sabre-toothed" and "conical-toothed" cats, respectively (Fig. 4.2; Werdelin 1996; Werdelin et al. 2008). Two phylogenies were recently published that include fossil taxa (Sakamoto and Ruta 2012; Christiansen 2013), but the relationships of most fossil felids are still poorly understood.

Machairodontinae Gill 1872
Smilodon Lund 1842
(Fig. 4.3; Tables 4.1 and 4.2)

Table 4.2 Diet and body mass of the South America carnivorans. The criterion for diet classification was taken from Prevosti et al. (2013). RGA: relative grinding area of lower carnassial (ml)

Taxa	RGA	Diet	Body mass (kg)	Commentaries
Smilodon populator	0	Hypercarnivore	290.00	Data from Prevosti et al. (2013)
Smilodon fatalis	0	Hypercarnivore	220.00	Data from Prevosti et al. (2013)
Smilodon gracilis	0	Hypercarnivore	77.50	Data from Prevosti et al. (2013)
Homotherium venezuelensis	0	Hypercarnivore	190.00	Data from Prevosti et al. (2013)
Panthera onca	0	Hypercarnivore	71.20	Data from Prevosti et al. (2013)
Puma concolor	0	Hypercarnivore	52.53	Data from Prevosti et al. (2013)
Herpailurus sp.	0	Hypercarnivore	5.20	Data from Prevosti et al. (2013)
Herpailurus yagouaroundi	0	Hypercarnivore	5.20	Data from Prevosti et al. (2013)
"Felis" vorohuensis	0	Hypercarnivore	3.77	Data from Prevosti et al. (2013)
Leopardus colocolo	0	Hypercarnivore	4.92	Data from Prevosti et al. (2013)
Leopardus geoffroyi	0	Hypercarnivore	2.89	Data from Prevosti et al. (2013)
Leopardus jacobita	0	Hypercarnivore	9.17	Data from Prevosti et al. (2013)
Leopardus guigna	0	Hypercarnivore	2.20	Data from Prevosti et al. (2013)
Leopardus pardalis	0	Hypercarnivore	10.08	Data from Prevosti et al. (2013)
Leopardus tigrinus	0	Hypercarnivore	3.00	Data from Prevosti et al. (2013)
Leopardus wiedii	0	Hypercarnivore	2.85	Data from Prevosti et al. (2013)
Lycalopex gymnocercus	0.54	Mesocarnivore	4.98	Data from Prevosti et al. (2013)
Lycalopex cultridens	0.55	Mesocarnivore	3.33	Data from Prevosti et al. (2013)
Lycalopex sp.	0.56	Mesocarnivore	4.20	Data from Prevosti et al. (2013)
Lycalopex culpaeus	0.49	Mesocarnivore	7.28	Data from Prevosti et al. (2013)
Lycalopex griseus	0.55	Mesocarnivore	3.33	Data from Prevosti et al. (2013)

(continued)

Table 4.2 (continued)

Taxa	RGA	Diet	Body mass (kg)	Commentaries
Lycalopex vetulus	0.68	Omnivore	3.35	Data from Prevosti et al. (2013)
Lycalopex fulvipes	0.59	Mesocarnivore	2.71	Data from Prevosti et al. (2013)
Lycalopex sechurae	0.60	Omnivore	3.60	Data from Prevosti et al. (2013)
Lycalopex ensenadensis	0.59	Omnivore	4.98	Data from Prevosti et al. (2013)
Cerdocyon thous	0.58	Hypercarnivore	5.70	Data from Prevosti et al. (2013)
Atelocynus microtis	0.56	Mesocarnivore	9.50	Data from Prevosti et al. (2013)
Chrysocyon brachyurus	0.51	Omnivore	25.00	Data from Prevosti et al. (2013)
Speothos venaticus	0.45	Hypercarnivore	6.50	Data from Prevosti et al. (2013)
Speothos pacivorus	0.45	Hypercarnivore	6.50	Data from Prevosti et al. (2013)
Dusicyon australis	0.48	Mesocarnivore	11.00	Data from Prevosti et al. (2013)
Dusicyon avus	0.48	Mesocarnivore	14.65	Data from Prevosti et al. (2013)
Theriodictis platensis	0.40	Hypercarnivore	36.00	Data from Prevosti et al. (2013)
"Canis" gezi	0.38	Hypercarnivore	36.00	Data from Prevosti et al. (2013)
Protocyon scagliorum	0.40	Hypercarnivore	25.00	Data from Prevosti et al. (2013)
Protocyon troglodytes	0.40	Hypercarnivore	25.00	Data from Prevosti et al. (2013)
Protocyon tarijensis	0.45	Hypercarnivore	30.00	Data from Prevosti et al. (2013)
Canis dirus	0.44	Hypercarnivore	51.00	Data from Prevosti et al. (2013)
Canis nehringi	0.40	Hypercarnivore	34.54	Data from Prevosti (2006b)
Canis familiaris	0.49	Omnivore	15.00	Data from Prevosti et al. (2013)
Urocyon cinereoargenteus	0.62	Omnivore	3.65	Data from Prevosti et al. (2013)
Urocyon sp	0.62	Omnivore	3.65	Data from Prevosti et al. (2013)
Arctotherium angustidens	0.79	Omnivore	900.00	Data from Prevosti et al. (2013)

(continued)

Table 4.2 (continued)

Taxa	RGA	Diet	Body mass (kg)	Commentaries
Arctotherium wingei	0.66	Omnivore	250.00	Data from Prevosti et al. (2013)
Arctotherium vetustum	0.66	Omnivore	300.00	Data from Prevosti et al. (2013)
Arctotherium tarijense	0.71	Omnivore	500.00	Data from Prevosti et al. (2013)
Arctotherium bonariense	0.69	Omnivore	600.00	Data from Prevosti et al. (2013)
Tremarctos ornatus	0.73	Omnivore	175.00	Data from Prevosti et al. (2013)
Mustela frenata	0.36	Hypercarnivore	0.12	Data from Prevosti et al. (2013)
Mustela africana	0.41	Hypercarnivore	0.19	Data from Prevosti et al. (2013)
Mustela felipei	0.41	Hypercarnivore	0.14	Data from Prevosti et al. (2013)
Eira barbara	0.46	Mesocarnivore	4.23	Data from Prevosti et al. (2013)
Galictis cuja	0.35	Hypercarnivore	1.90	Data from Prevosti et al. (2013)
Galictis sorgentinii	0.55	Hypercarnivore	2.12	Data from Prevosti et al. (2013)
Galictis vittata	0.47	Hypercarnivore	2.34	Data from Prevosti et al. (2013)
Galictis hennigi	0.35	Hypercarnivore	2.12	Data from Prevosti et al. (2013)
Galictis sp.	0.35	Hypercarnivore	2.12	Data from Prevosti et al. (2013)
Lyncodon bosei	0.47	Hypercarnivore	0.23	Data from Prevosti et al. (2013)
Lyncodon patagonicus	0.44	Hypercarnivore	0.23	Data from Prevosti et al. (2013)
Stipanicicia pettorutii	0.41	Hypercarnivore	2.12	Data from Prevosti et al. (2013)
Lontra longicaudis	0.87	Hypercarnivore	10.33	Data from Prevosti et al. (2013)
Lontra provocax	0.68	Hypercarnivore	7.50	Data from Prevosti et al. (2013)
Lontra felina	0.86	Hypercarnivore	4.40	Data from Prevosti et al. (2013)
Pteronura brasiliensis	0.70	Hypercarnivore	26.00	Data from Prevosti et al. (2013)
Conepatus chinga	1.11	Omnivore	1.92	Data from Prevosti et al. (2013)

(continued)

Table 4.2 (continued)

Taxa	RGA	Diet	Body mass (kg)	Commentaries
Conepatus mercedensis	1.22	Omnivore	2.36	Data from Prevosti et al. (2013)
Conepatus talarae	1.27	Omnivore	2.36	Data from Prevosti et al. (2013)
Conepatus semistriatus	1.13	Omnivore	4.00	Data from Prevosti et al. (2013)
Conepatus primaevus	1.07	Omnivore	4.00	RGA: this work; Body mass of *C. semistriatus*
Nasua nasua	1.23	Omnivore	5.00	Data from Prevosti et al. (2013)
Nasua narica	1.23	Omnivore	5.00	Data from Prevosti et al. (2013)
Nasuella olivacea	1.23	Omnivore	4.00	Data from Prevosti et al. (2013)
Procyon cancrivorus	0.87	Omnivore	9.00	Data from Prevosti et al. (2013)
Potos flavus	2.08	Omnivore	3.00	Data from Prevosti et al. (2013)
Potos sp.	2.08	Omnivore	3.00	Data from Prevosti et al. (2013)
Bassaricyon gabbi	1.34	Omnivore	1.20	Data from Prevosti et al. (2013)
Bassaricyon bedarddi	1.34	Omnivore	1.20	Data from Prevosti et al. (2013)
Bassaricyon alleni	1.34	Omnivore	1.20	Data from Prevosti et al. (2013)
Cyonasua argentina	0.88	Omnivore	12.97	This work
Cyonasua pascuali	0.88	Omnivore	11.37	This work
Cyonasua groeberi	0.95	Omnivore	14.73	This work
Cyonasua brevirostris	0.95	Omnivore	14.19	This work
Cyonasua clausa	0.86	Omnivore	15.32	This work
Cyonasua lutaria	0.77	Omnivore	14.26	This work
Cyonasua meranii	0.88	Omnivore	13.49	This work
Chapalmalania ortognatha	0.79	Omnivore	88.13	This work
Chapalmalania altaefrontis	0.75	Omnivore	88.13	This work

Species in South America: *Smilodon fatalis* Leidy 1868, *S. gracilis* Cope 1880, and *S. populator* Lund 1842.

Temporal and geographic distribution: In South America, *Smilodon* is recorded from the Ensenadan to the Lujanian, from Tierra del Fuego in Chile to Venezuela, covering almost all of South America (Berta 1985; Kurtén and Werdelin 1990; Cartelle 1999; Ubilla and Perea 1999; Hadler Rodríguez et al. 2004; Rincón 2006; Prevosti and Pomi 2007; Prevosti et al. 2013; Fariña et al. 2014; Lindsey and Seymour 2015).

Paleoecology: *Smilodon* was a very large hypercarnivore that predated on large mammals (Christiansen 2008; Slater and Van Valkenburgh 2008; Prevosti et al. 2010), using a stalk and ambush strategy (Akersten 1985; Van Valkenburgh and Hertel 1998; Coltrain et al. 2004; Prevosti and Vizcaíno 2006; Wroe et al. 2013; Prevosti and Martin 2013; Morales and Giannini 2014). *Smilodon* may have been a social predator (Carbone et al. 2009; Van Valkenburgh et al. 2009; Bocherens et al. 2016; but see McCall et al. 2003; Kiffner 2009). Using stable isotopes, Coltrain et al. (2004) suggested that *S. fatalis* hunted ruminants in North America (Rancho La Brea), while Prevosti and Martin (2013) and Bocherens et al. (2016) stated that *S. populator* consumed horses, ground sloths, and camelids in southern Patagonia and large mammals from open habits (e.g., macrauchenids and ground sloths) in the Pampean Region, respectively. The estimated body mass is between 55 kg and 100 kg for *S. gracilis*, 160 kg–280 kg for *S. fatalis*, and 220 kg–400 kg for *S. populator* (Christiansen and Harris 2005; Prevosti and Vizcaíno 2006; Prevosti and Martin 2013). The ambush strategy would have resulted in a bite to the throat of the prey powered by the massive jaw and neck muscles (Akersten 1985; Andersson et al. 2011). The robust forelimbs were important to fix the prey during the bite, and would have contributed to the bite power, generating a class 1 lever (McHenry et al. 2007; Meachen-Samuels and Van Valkenburgh 2009; Brown 2014).

Comments: *Smilodon* initially appeared in the late Pliocene of North America with the species *Smilodon gracilis* (Berta 1985, 1995). The immigration of this genus to South America is obscured by the limited fossil record of Central America and northern South America (Rincón et al. 2011). The characters used to separate *S. fatalis* from *S. populator* are variable, and some specimens have a mix of them, indicating that the genus needs a systematic review.

Recently, Chimento (2016) questioned the synonymy between *Smilodontidium riggii* Kraglievich 1948, and *Smilodon* (see Prevosti and Pomi 2007) and considered it a pantherine. This conclusion, however, was based on only two specimens of *S. populator*, without considering the intraspecific variation (or intrageneric in this case) among *Smilodon* and other felids. They utilized a selection of characters of unproved systematic value, ignoring other relevant features (e.g., length of the head of the astragalus; depth of the medial groove for the astragalar trochlea on the distal articular tibial facet in anterior view, robustness of the tibia diaphysis) that are shared with *Smilodon* (Prevosti and Pomi 2007).

Homotherium Fabrini 1890
(Fig. 4.4; Tables 4.1 and 4.2)

Fig. 4.4 Lateral view of the skull of *Homotherium venezuelensis* (IVIC OR 1352, Holotype)
Scale = 5 cm

5 cm

Species in South America: *Homotherium venezuelensis* Rincón, Prevosti and Parras 2011.

Temporal and geographic distribution: The South American species is restricted to the early–Middle Pleistocene of eastern Venezuela (Rincón et al. 2011).

Paleoecology: The morphology of *H. venezuelensis* is similar to its Holarctic congeners, and indicates that it had a body mass of ca. 190 kg and preyed on large mammals (Van Valkenburgh and Hertel 1998; Antón and Galobart 1999).

Comments: An incomplete mandible from Uruguay, with imprecise age, was assigned to cf. *Xenosmilus* (Mones and Rinderknecht 2004). The specimen clearly represents Homotheriini, but its generic attribution is uncertain (Rincón et al. 2011).

Felinae Fischer 1817
Panthera Oken 1816
(Fig. 4.5a; Tables 4.1 and 4.2)

Species in South America: *Panthera onca* (Linnaeus 1758).

Temporal and geographic distribution: In South America, *Panthera* is recorded from the Ensenadan to Present, and from Tierra del Fuego in Chile to northern South America (Cabrera 1934; Ochsenius and Gruhn 1979; Seymour 1983; Hoffstetter 1986; Cartelle 1999; Ubilla and Perea 1999; Pomi and Prevosti 2005; Martin 2013; Prevosti and Martin 2013).

Paleoecology: Living jaguars are the largest predators (30 kg–120 kg) in South America, hunting large- and medium-sized mammals (Wilson and Mittermeier 2009). During the Pleistocene, a much larger jaguar, with a body mass of about 190 kg, was present in Peru, Argentina, and southern Chile (Cabrera 1934; Seymour 1983; Pomi and Prevosti 2005). The Patagonian Jaguar has been given different names (the most common is *Panthera onca mesembrina* coined by Angel Cabrera 1934). A recent paleoecological study combining morphology, stable isotopes, and taphonomy demonstrated that the Patagonian Jaguar was able to hunt horses, camelids, and ground sloths (*Mylodon*) (Martin 2013; Prevosti and Martin 2013). A similar conclusion can be inferred for the Late Pleistocene jaguars found in the Pampean Region (Prevosti and Vizcaíno 2006). Recent isotopic data indicate

Fig. 4.5 Lateral view of the skull of a juvenil specimen of *Panthera onca mesembrina* (BM M20893) (**a**), lateral view of the skull (**b**), and mandible (**c**) of *Puma concolor* (MMP 1476 M). Scale = 5 cm

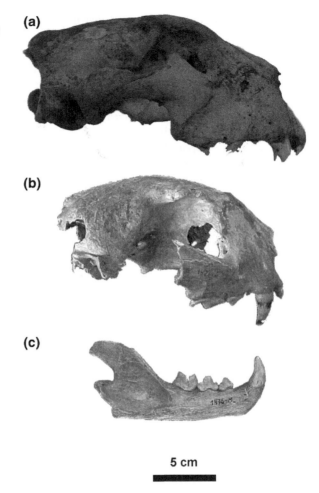

(a)

(b)

(c)

5 cm

that it hunted macrauchenids, ground sloths, chinchillids, and equids; the first two groups in lower proportion, and the last two in larger proportion than in *Smilodon populator* and *Protocyon troglodytes* (Lund 1838) (Bocherens et al. 2016). This implies lower competition between these large felids than between *Smilodon fatalis* and *Panthera atrox* (Leidy 1853) from the Late Pleistocene of North America, probably related to the larger size of the South American species (Bocherens et al. 2016).

Comments: The oldest records of *Panthera onca* in South America are from the early–Middle Pleistocene (Ensenadan, 1.8–0.5 Ma) of the Pampean Region (Berman 1994; Ubilla and Perea 1999; Soibelzon and Prevosti 2007; Prevosti and Soibelzon 2012). In North America, the oldest records are of similar age (middle Irvingtonian, 1–0.6 Ma; Seymour 1993; Turner and Antón 1996; Woodburne 2004). Some authors have suggested that jaguars originated in Africa (Hemmer et al. 2010), while others suggested that the Old World, early Pleistocene taxon (e.g., *Panthera gombaszoegensis* (Kretzoi 1858)) was related to or should even be included in *P. onca* (Seymour 1993; Turner and Antón 1996). These hypotheses are based only on a comparative qualitative approach and should be tested using cladistic methodology. With the same approach mentioned above, and claiming the authority principle, Chimento (2016) considered that *"Felis" longifrons* Burmeister 1866 is not a synonym of *P. onca* (see Pomi and Prevosti 2005) but a different taxon. Unfortunately, omission of a complete discussion of previous systematic arguments, that support the synonymy, do not help to resolve the systematic position of this fossil (Pomi and Prevosti 2005).

Puma Jardine 1834

(Fig. 4.5b, c; Tables 4.1 and 4.2)

Species in South America: *Puma concolor* (Linnaeus 1771).

Temporal and geographic distribution: This species has been present in South America since the Ensenadan (1.8–0.5 Ma). The fossil record suggests a broad distribution in the continent (e.g., Hoffstetter 1952; Berta and Marshall 1978; Berman 1994; Ubilla and Perea 1999; Cartelle 1999; Shockey et al. 2009; Soibelzon and Prevosti 2007; Prevosti and Martin 2013), as is the case at present, with a species range covering all SA (Wilson and Mittermeier 2009). An excellent skull of *Puma concolor* (MMP 1476 M) found in Ensenadan levels of the sea-cliffs placed north of Mar del Plata city (Buenos Aires, Argentina) is shown in Fig. 4.5.

Paleoecology: *Puma* is one of the largest predators (34 kg–120 kg) in SA today and is distributed over a wide range of environments, such as cold steppes and tropical rainforests, where they hunt medium- and large-sized mammals, including domestic animals (e.g., cows and horses; Wilson and Mittermeier 2009). Larger pumas, with a body mass of about 140 kg, were recorded in Southern Patagonia as well as elsewhere (Prevosti and Vizcaíno 2006) during the Late Pleistocene, where they were able to prey on native horses (*Hippidium* and *Equus* (*Amerhippus*)) and camelids (Prevosti and Martin 2013).

Comments: Pumas, together with the small jaguarundi (*Herpailurus* Severtzov 1858), are related to the cheetah (*Acinonyx*) and represent another felid lineage that invaded SA during the Pleistocene. Its phylogenetic relationship, and the inclusion of the North American and Old Word fossil species in the same genus, was suggested by morphological and morphometric analyzes (Seymour 1999; Hemmer et al. 2004; Madurell Malapeira et al. 2010) but still has no cladistic support. The fossil record of *Puma concolor* is older in SA than in NA, since the oldest record in North America is 0.4 Ma (Van Valkenburgh et al. 1990). Living NA populations apparently constitute a re-migration from SA that occurred in the Late Pleistocene (10 ka; Culver et al. 2000).

Herpailurus Severtzov 1858

(Tables 4.1 and 4.2)

Species in South America: *Herpailurus yagouaroundi* (E. Geoffroy Saint-Hilaire 1803).

Temporal and geographic distribution: *H. yagouaroundi* is today distributed from southern NA to northern Patagonia, mainly in forested areas (Wilson and Mittermeier 2009). Its fossil record in SA is very incomplete and limited to the "Lujanian" of Minas Gerais and Bahia, Brazil (10–200 ka) and cf. *Herpailurus* from the Bonaerian (0.5–0.125 Ma) of the Pampean Region, Argentina (Prevosti 2006a). It is also mentioned in a faunal list of the Lujanian of Piauí; Brazil (48.5–12 ka BP; Guérin et al. 1993).

Paleoecology: Extrapolation of the ecological habits of the living species suggests that fossil *Herpailurus* were small predators (3 kg–7.6 kg) of small mammals (mainly rodents) (Wilson and Mittermeier 2009).

Comments: The long genetic separation of *Puma* and *Herpailurus* and their large morphological differences support the view that *Herpailurus* independently invaded South America after the establishment of the Panama Bridge (Prevosti 2006a; Rincón et al. 2011; Prevosti and Soibelzon 2012). Recently, Chimento et al. (2014) suggested that *"Felis" pumoides* Castellanos 1958, a species previously referred to *Puma concolor* (Berman 1994), is valid. The authors included the taxon in *Herpailurus* and assigned it to the Pliocene. However, the characters used to justify their taxonomic attribution are variable among felids or taxonomically irrelevant (Seymour 1999; Prevosti *pers. obs.*). For example, the minor palatine foramen has the same development in some of the living specimens of *P. concolor* (e.g., MACN 13054) studied by Chimento et al. (2014). Instead, the size and morphology of this fossil are close to the living *Puma*, as indicated by Berman (1994). At least one feature is different from the living specimens of *Puma concolor* studied, the shape of the P3 (see Castellanos 1958), which is narrower and has a less developed lingual cingulum in its distal portion (Prevosti *pers. obs.*). Importantly, the author did not provide evidence in support of a much earlier age than the accepted Ensenadan or younger age.

Leopardus Gray 1842
(Tables 4.1 and 4.2)

Species in South America: *Leopardus colocolo* (Molina 1782), *L. geoffroyi* (d'Orbigny and Gervais 1844), *L. guigna* (Molina 1782), *L. jacobita* (Cornalia 1782), *L. pardalis* (Linnaeus 1758), *L. tigrinus* (Schreber 1775), *L. wiedii* (Schinz 1821), and *"Felis" vorohuensis* Berta 1983.

Temporal and geographic distribution: *Leopardus* is dispersed across the continent. *Leopardus tigrinus*, *L. wiedii*, and *L. pardalis* are also present in Central America and southern NA (Wilson and Mittermeier 2009). The oldest records in SA are *L. colocolo* and *"Felis" vorohuensis* from the late Ensenadan (0.78–0.5 Ma) of the Pampean Region (Prevosti 2006a). *Leopardus tigrinus*, *L. wiedii,* and *L. pardalis* are recorded from the Late Pleistocene–early Holocene ("Lujanian") of Minas Gerais, Brazil, *L. pardalis* from the "Lujanian" of Bahia, Brazil, the Lujanian of Talara, Peru (ca. 13 ka BP), and the late Holocene of Ecuador (Cartelle 1999;

Prevosti 2006a; Stahl 2003). *Leopardus wiedii* and *L. pardalis* have also found in the Late Pleistocene of southern USA (Werdelin 1985; Seymour 1999; Prevosti 2006a). Remains of *L. geoffroyi* are from the Holocene, whereas the living species *L. guigna* and *L. jacobita* do not have a definitive fossil record (Prevosti 2006a). Linares (1998) mentioned the presence of *L. pardalis* in archaeological sites of Margarita Island, Central and Oriental Cordilleras, and *L. tigrinus* in the Central Cordillera, and the San Pedro Archipelago in Venezuela, but he did not figure or describe these remains, and the attributions were not justified. *Leopardus pardalis* was recently reported from the Sopas Formation (27–58 ka BP) in Uruguay (Perea et al. 2015). *Leopardus colocolo* is also cited for the Platan of Buenos Aires, Argentina, and southern Patagonia in Chile (Clutton Brock 1988; Quintana 2001; Alvarez 2009; see also Prevosti 2006a).

Paleoecology: According to the ecology of the living species, it is possible that the fossil taxa were small to mid-size predators (1.3 kg–15 kg) of small mammals, mainly rodents (Wilson and Mittermeier 2009).

Comments: Two alternative hypotheses can be drawn for *Leopardus*. The genus invaded SA after the development of the Panama Bridge, where it was diversified into several species in relation to different biogeographic areas and prey sizes, later re-invading Central and North America. Or several species might have invaded SA independently (Prevosti 2006a; Eizirik 2012; Prevosti and Soibelzon 2012; Prevosti and Pardiñas, in press). The last hypothesis could be supported by the fact that *L. pardalis* and *L. wiedii* are sister taxa that have been separated from other species of the genus for as much as 2 and 4.25 Ma (see Johnson et al. 2006) and have Late Pleistocene fossils in southern NA (Werdelin 1985; Seymour 1999). Diversity in specialization to different environments was an important factor in the evolution of this genus (Prevosti and Pardiñas, in press).

Recently, *L. guttulus* (Hensel 1872) was recognized as a full species separate from *L. tigrinus* based on genetic data (Trigo et al. 2013); this hypothesis should be also evaluated with morphological data and tested with more evidence.

Caniformia Kretozoi 1938
Canidae Fischer 1817

Living canids (Caninae) have quite generalized habits. They retain a more complete dentition than do Felidae, a long rostrum, and cursorial locomotion (Ewer 1973; Wang and Tedford 2008; Tedford et al. 2009). However, extinct canid groups demonstrate much larger morphological disparity and diverse ecological habits (Wang and Tedford 2008; Tedford et al. 2009). The family has a long history in NA, where they are recorded from the late Eocene (36 Ma) onwards. Later, between the middle and late Miocene (16 and 7 Ma), they reached the Old World (Wang and Tedford 2008; Tedford et al. 2009). The family experienced several radiations in NA that resulted in wide morphological disparity and ecomorphs, with omnivorous, insectivorous, scavenging and hypercarnivorous diets, and scansorial and cursorial locomotor types (Wang 1993; Wang et al. 1999). Canids invaded South America in different waves during the Pleistocene, and their diverse specialization to different environments was an important feature of their radiation (Prevosti 2010; Eizirik 2012; Prevosti and Soibelzon 2012; Prevosti and Pardiñas, in press; Moura Bubadué et al. 2015). *Atelocynus* is the only living genus that lacks confirmed fossils.

Lycalopex Burmeister 1854
(Tables 4.1 and 4.2)

Species in South America: *Lycalopex culpaeus* (Molina 1782), *L. cultridens* (Gervais and Ameghino 1880), *L. ensenadensis* (Ameghino 1888), *L. fulvipes* (Martin 1837), *L. griseus* (Gray 1837), *L. gymnocerus* (Fischer 1814), *L. sechurae* Thomas 1900, and *L. vetulus* (Lund 1842).

Temporal and geographic distribution: The oldest record of *Lycalopex* is *L. cultridens* from the Vorohuean (late Pliocene) of Buenos Aires, Argentina. This species disappeared in the Ensenadan (Berman 1994; Soibelzon and Prevosti 2007, 2013). Also in Buenos Aires, a fox similar to but larger than *L. gymnocercus* was found in Sanandresian levels (late Pliocene–earliest Pleistocene). Fossil remains confidently assigned to *L. gymnocercus* are known since the Ensenadan (early–Middle Pleistocene) (Berman 1994; Soibelzon and Prevosti 2007, 2013). *Lycalopex gymnocercus* is also found in the Lujanian of Entre Ríos, Argentina; Salto, Uruguay; "uncertain" levels in Tarija, Bolivia; the "Lujanian" of Minas Gerais, Brazil (Berta 1987; Soibelzon and Prevosti 2007, 2013), and the Platan of Buenos Aires and Santiago del Estero, Argentina (Ameghino 1889; Bonomo 2005; Alvarez 2009; Quintana 2001; del Papa 2012). *Lycalopex ensenadensis* is to date exclusively from Ensenadan and Lujanian levels in Buenos Aires, (Ramirez and Prevosti 2014). *Lycalopex culpaeus* is present in Lujanian–Platan sites of Southern Chile and Patagonia in Argentina, the "Lujanian" of La Carolina and Punin, Ecuador; and the Lujanian of Huánuco and Junín, Peru (Caviglia 1986; Clutton-Brock 1988; Massoia 1992; Trejo and Jackson 1998; Soibelzon and Prevosti 2007, 2013; Amorosi and Prevosti 2008; Shockey et al. 2009; Méndez et al. 2014). Berman (1994) described a specimen from the Ensenadan of Buenos Aires as *L. culpaeus,* but the specimen is very fragmentary and does not preserve the diagnostic features of the species. *Lycalopex sechurae* was found in Talara, Peru (Lujanian, 13–14 ka BP), and La Carolina, Ecuador ("Lujanian"), *L. vetulus* in the "Lujanian" of Minas Gerais, Brazil, and *L. griseus* in the Holocene of Patagonia, and possibly also Córdoba and Santiago del Estero, Argentina (Ameghino 1889; Kraglievich and Rusconi 1931; Berta 1987; Clutton-Brock 1988; Massoia 1992; Soibelzon and Prevosti 2007, 2013; Amorosi and Prevosti 2008; Méndez et al. 2014). *Lycalopex sechurae* and *L. culpaeus* are also included in faunal lists of Holocene archaeological sites in Ecuador (Stahl and Athens 2001; Stahl 2003). The presence of *L. griseus* and *L. culpaeus* in the Quaternary of Buenos Aires, Argentina, is not corroborated (but see comments for the taxonomic status of *L. gymnocercus* and *L. griseus*).

Paleoecology: *Lycalopex* is a small to middle-sized canid (1.8 kg–13.8 kg) with mesocarnivorous-omnivorous habits. In particular, *L. culpaeus* is the more carnivorous species, hunting rodents and hares (Wilson and Mittermeier 2009).

Comments: The genus *Lycalopex* originated in SA, where it diversified into at least seven species, occupying different geographic areas and environments (Prevosti and Soibelzon 2012; Prevosti and Pardiñas, in press). Morphological studies suggest that *L.griseus* is a junior synonym of *L. gymnocercus* (Zunino et al. 1995; Prevosti et al. 2013), but this should be corrroborated with DNA studies and other sources of information. The extinct *L. cultridens* has a morphology and size intermediate between *L. gymnocercus* and *L. griseus* (Berman 1994), thus this fossil should be included in *L. gymnocercus* if this synonymy is accepted. *"Canis" peruanus* Nordenskiöld 1908 (Late Pleistocene of Peru; Nordenskiöld 1908) has a cranial and dental anatomy that agrees with *L. culpaeus*, a species also present in the region since the Late Pleistocene.

Cerdocyon Hamilton Smith 1839
(Tables 4.1 and 4.2)

Species in South America: *Cerdocyon thous* (Linnaeus 1766).

Temporal and geographic distribution: The recent distribution of this species covers most of north and central SA, across forested environments with the exception of Amazonia (Wilson and Mittermeier 2009). Few fossils are assigned to this genus, which is limited to the "Lujanian" of Minas Gerais and Bahia, Brazil, and the Lujanian of the Pampean Region, Argentina (Berta 1987; Cartelle 1999; Ramírez 2014).

Paleoecology: *Cerdocyon thous* is a small fox (4.5 kg–8.5 kg) with an omnivorous diet composed of fruits, insects, and small mammals (Wilson and Mittermeier 2009).

Comments: Some molecular studies suggest a SA origin for *Ce. thous* (Eizirik 2012). However, the long molecular divergence between *Cerdocyon* and other SA canids suggests that the genus could have differentiated outside South America (i.e., Central and/or North America; Prevosti 2010; Prevosti and Soibelzon 2012; Prevosti and Pardiñas, in press). But the supposed records of *Cerdocyon* in the NA Pliocene (Tedford et al. 2009) are incorrect because the specimens in question appear to be more closely related to Vulpini (Prevosti 2010).

Chrysocyon Hamilton Smith 1839
(Tables 4.1 and 4.2)

Species in South America: *Chrysocyon brachyurus* (Illiger 1815).

Temporal and geographic distribution: Extant populations *Ch. brachyurus* are present in northeastern Argentina, Uruguay, South and Central Brazil, Paraguay, and Bolivia, inhabiting grasslands, crop fields, "Cerrado" and other forested areas (Wilson and Mittermeier 2009). Fossil remains are known from the "Lujanian" of Minas Gerais, Brazil, and the Holocene of Buenos Aires and Entre Ríos, Argentina (Prevosti et al. 2009a). Other material comes from Tarija, Bolivia, but its stratigraphic location in the Ensenadan of that locality is doubtful (Prevosti et al. 2009a).

Paleoecology: *Chrysocyon brachyurus* is a large canid (20.5 kg–30 kg), with an omnivorous diet composed of fruits, insects, and rodents (Wilson and Mittermeier 2009). The functional relevance of its long legs is a matter of discussion. Some authors have suggested that the long legs are useful for locomotion in flooded areas, or alternatively for locomotion in grasslands, or for rapid pursuit of prey (Hildebrand 1952; 1954; Langguth 1975; Dietz 1985; Andersson 2004).

Comments: *Chrysocyon* has a poor fossil record in SA. The fact that *Ch. nearcticus* was described from the Pliocene of North America suggested that the genus originated in the northern Hemisphere (Wang and Tedford 2008; Tedford et al. 2009). A new phylogenetic analysis resulted in *Ch. nearcticus* Wang, Tedford, and Taylor 1999 being placed together with other SA canids, but not the sister taxon of *Ch. brachyurus*. This could be an artifact of the incompleteness of the fossils, indicating that more material is needed to test this hypothesis (Prevosti 2010). Additionally, if the phylogeny of Prevosti (2006b, 2010; see also Austin et al. 2013) is correct regarding the position of *Theriodictis? floridanus* (see below), the *Chrysocyon* clade could well have originated in North America. The presence of *Ch. brachyurus* in the Pampean Region during the Holocene, beyond the southern limit of its current distribution, was explained by the presence of warmer temperatures at that time (Prevosti et al. 2004).

Speothos Lund 1839

(Tables 4.1 and 4.2)

Species in South America: *Speothos venaticus* (Lund 1842) and the extinct *Sp. pacivorus* Lund 1842.

Temporal and geographic distribution: *Speothos* occurs today from eastern Panama to northeastern Argentina in humid forests (Berta 1984; Wilson and Mittermeier 2009). Fossils of *Sp. venaticus* and *Sp. pacivorus* are few and restricted to the "Lujanian" of Brazil and a mention of *Speothos* sp. in the faunal list of the late Holocene site of Loma Alta, Ecuador (Stahl 2003).

Paleoecology: *Speothos venaticus* is a small canid (5 kg–8 kg.) with a long body, short legs, short tail, and a reduced and specialized dentition associated with a hypercarnivorous diet (Kraglievich 1930; Hildebrand 1952, 1954; Langguth 1980; Berta 1984; Prevosti 2006b). A good swimmer, it hunts large rodents (e.g., *Dasyprocta* sp.) and armadillos (e.g., *Dasypus* sp.), that it can pursue in their burrows or in the water (Cabrera and Yepez 1940; Wilson and Mittermeier 2009; Lima et al. 2009).

Comments: The putative fossil species *Sp. pacivorus* could be a junior synonym of *Sp. venaticus*, and the differences explained by intraspecific variation. Lima et al. (2009) suggested that the anatomy of *Speothos* evolved to hunt mammals inside burrows, while Hildebrand (1954) argued that the short legs are a specialization for the thick undergrowth bordering jungle streams along which it hunts.

Dusicyon Hamilton Smith 1839

(Fig. 4.6b–c, Tables 4.1 and 4.2)

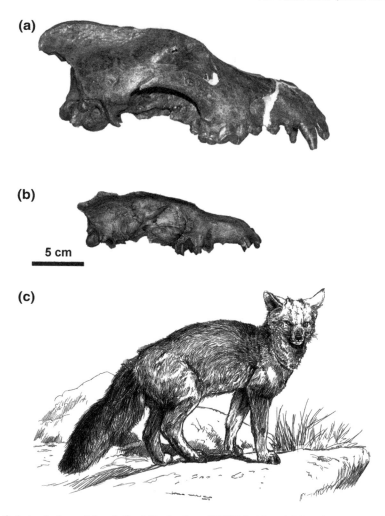

Fig. 4.6 Lateral view of the skull of *Canis dirus* (VF WN) (**a**) and "*Canis*" *peruanus* (NHRM M1952, Holotype) (**b**), and life reconstruction of *Dusicyon avus* (**c**). Scale = 5 cm

Species in South America: *Dusicyon australis* (Kerr 1792), *D. avus* (Burmeister 1866).

Temporal and geographic distribution: *D. australis* was endemic to the Malvinas (= Falkland) islands and was extirpated in the nineteenth century by human hunting. The continental *D. avus* lived in southern Brazil, Uruguay, and the Pampean and Patagonian regions of Argentina during the Late Pleistocene and Holocene (Prevosti et al. 2011, 2015).

Paleoecology: The body mass of *D. avus* was estimated to lie between 13 kg–17 kg. Its diet was apparently more carnivorous than modern SA foxes, which included hunting rodents, armadillos, and juvenile ungulates (e.g., camelids). Stable

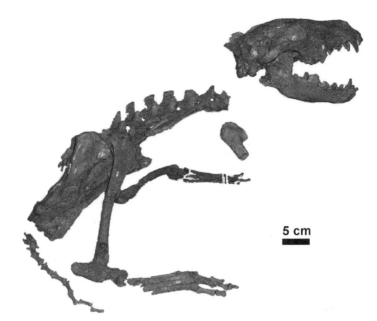

Fig. 4.7 Mounted skeleton of *Theriodictis platensis* (MPS 2). Scale = 5 cm

isotope analysis indicates that *D. avus* could scavenge megammals during the latest Pleistocene and early Holocene in Patagonia (i.e., *Mylodon* sp.; Prevosti and Vizcaíno 2006; Prevosti and Martin 2013).

Comments: Genetic and morphological data indicate that *Dusicyon* is the sister taxon of *Chrysocyon* and is part of the clade that also included *Speothos*, *Protocyon* Giebel 1855, and *Theriodictis* Mercerat 1891(Austin et al. 2013). Available ecological information indicates that *D. australis* hunted birds and ate sea mammals (Cabrera and Yepez 1940). *Dusicyon avus* became extinct around 400–500 years ago, probably by a combination of environmental changes and human impact (Prevosti et al. 2015)

Theriodictis Mercerat 1891
(Fig. 4.7, Tables 4.1 and 4.2)

Species in South America: *Theriodictis platensis* Mercerat 1891.

Temporal and geographic distribution: Ensenadan of the Pampean Region (Prevosti 2006b; Prevosti et al. 2009a).

Paleoecology: *Theriodictis platensis* was one of the largest hypercarnivorous canids from SA (30 kg–50 kg), able to hunt medium- and large-sized mammals (Prevosti and Palmqvist 2001; Prevosti 2006b). The postcranial anatomy suggests that it was cursorial, similar to *Lycaon pictus* (Temminck 1820) (Prevosti 2006b).

Comments: Tedford et al. (2009) tentatively identified the species *Theriodictis*? *floridanus* Wang et al. 1999 from the late Pliocene–early Pleistocene of NA. In a recent phylogenetic analysis, this species was placed as the sister taxon of

Protocyon + *Theriodictis* + *Speothos* + "*Canis*" *gezi* Kraglievich 1928, which may suggest that the origin of this clade was in North America (Prevosti 2010). Unfortunately, *Theriodictis*? *floridanus* is only known by an incomplete mandible and isolated teeth and more complete material is needed to evaluate this hypothesis.

"*Canis*" *gezi* Kraglievich 1928
(Tables 4.1 and 4.2)

Species in South America: "*Canis*" *gezi* Kraglievich 1928.

Temporal and geographic distribution: Ensenadan of the Pampean Region, Argentina (Kraglievich 1928; Berta 1989; Prevosti 2006b; Prevosti et al. 2009a).

Paleoecology: "*Canis*" *gezi* was a large (36 kg–37 kg) hypercarnivorous canid, able to prey on animals between 40 kg–200 kg, or larger including cervids, equids, camelids, peccaries, mesotheriids, large rodents, and armadillos (Prevosti 2006b). The few postcranial remains indicate that it had cursorial habits similar to the living *Canis lupus* or *Lycaon pictus* (Prevosti 2006b).

Comments: Cladistic analysis demonstrated that "*Canis*" *gezi* is not related to *Canis*, but to the clade of *Theriodictis, Protocyon,* and *Speothos* (Prevosti 2006b, 2010). "*Canis*" *gezi* is very similar to *Theriodictis*, but has plesiomorphic traits in the dentition, like the presence of a metaconid in the lower carnassial and a more developed hypoconid in the first upper molar (Prevosti 2006b). Since this taxon is known by two incomplete specimens, it is difficult to conclude if it represents another canid or if it could be part of the variation of *Th. platensis*.

Protocyon Giebel 1855
(Tables 4.1 and 4.2)

Species in South America: *Protocyon scagliorum* JL Kraglievich 1952, *Pr. tarijensis* (Ameghino 1902), and *Pr. troglodytes* (Lund 1838).

Temporal and geographic distribution: *Protocyon scagliorum* is known from the Ensenadan (0.78–0.5 Ma BP) of the Pampean Region, Argentina, while *P. tarijensis* is from Tarija, Bolivia, between 1 Ma and ca. 27 ka BP (Prevosti et al. 2009a). *Protocyon troglodytes* is widely distributed during the Late Pleistocene, including Argentina, Uruguay, Bolivia, Brazil, Ecuador, and Venezuela; older records of this species are questionable (Prevosti 2006b; Prevosti and Rincón 2007; Prevosti et al. 2009a; Lindsey and Seymour 2015). The genus *Protocyon* was mentioned from the Middle Pleistocene tar pit of Orocual in eastern Venezuela (Rincón et al. 2009).

Paleoecology: The genus *Protocyon* includes large canids (20 kg–40 kg), with cursorial habits, able to hunt medium- to large-sized mammals (e.g., equids, camelids, deer, tayassuids, and juveniles of megammals) (Cartelle and Langguth 1999; Prevosti 2006b; Prevosti et al. 2009b). Stable isotope analyses indicate that *P. troglodytes* was a hypercarnivore, with equid and megammals carrion probably forming part of its diet (Prevosti and Schubert 2013). A new analysis that includes a larger sample and Bayesian methods (Bocherens et al. 2016) shows that *P. troglodytes* mainly ate macrauchenids and ground sloths, and that had an important overlap in diet with *Smilodon populator*. The exploitation of the same resource could imply an important level of competition between these species.

Comments: *Theriodictis tarijensis* was transferred to *Protocyon* on the basis of the results of phylogenetic analyses (Prevosti 2006b; 2010). A specimen of *Pr. troglodytes* from the Pampean Region provided an age of 17.34 ka BP by AMS 14C (Prevosti and Schubert 2013).

Canis Linnaeus 1758

(Fig. 4.6a; Tables 4.1 and 4.2)

Species in South America: *Canis dirus* Leidy 1858, *C. familiaris* Linnaeus 1758 (domestic dog), and *C. nehringi* (Ameghino 1902).

Temporal and geographic distribution: *Canis dirus* and *C. nehringi* have been found in the latest Pleistocene of Bolivia, Peru, Venezuela, and Argentina (Prevosti 2006b; Prevosti and Rincón 2007; Prevosti et al. 2009a). Domestic dogs are recorded in the continent since the Holocene (Prates et al. 2010).

Paleoecology: *Canis nehringi* and *C. dirus* were one of the larger SA canids (30 kg–38 kg and 30 kg–70 kg, respectively). Horses, cervids, camelids, large rodents, tayassuids, and juveniles of megammals could have been hunted by these canids (Prevosti 2006b; Prevosti and Vizcaíno 2006 and references therein).

Comments: *Canis nehringi* and *C. dirus* were very similar (Berta 1989), possible the same species (Prevosti 2006b). The late presence of *Canis nehringi* and *C. dirus* in South America, and the long history of the last species in NA (Tedford et al. 2009), suggest a younger (Late Pleistocene) immigration to the southern continent (Prevosti 2006b, 2010; Prevosti and Rincón 2007).

Urocyon Baird 1857

(Tables 4.1 and 4.2)

Species in South America: *Urocyon cinereoargenteus* (Schreber 1775).

Temporal and geographic distribution: *Urocyon cinereoargenteus* currently inhabits the northwestern part of SA (Colombia and Venezuela), Central America, and a major part of NA. Its fossil record in SA is limited to the Late Pleistocene of western Venezuela (Prevosti and Rincón 2007).

Paleoecology: *Urocyon cinereoargenteus* is a small canid (2 kg–5.5 kg) with omnivorous diet, including fruits, insects, and small vertebrates, and with scansorial capabilities (Hildebrand 1954; Wilson and Mittermeier 2009).

Comments: The genus *Urocyon* has an extensive record in NA (late Miocene–Present; Wang and Tedford 2008; Tedford et al. 2009), unlike SA with records only as old as the Late Pleistocene (Prevosti and Rincón 2007) that point to a recent immigration.

Ursidae Gray 1825

The bear family includes large terrestrial carnivores with stout legs and fully plantigrade hind feet. Dietary variation in the group encompasses mostly omnivorous (spectacled bear, *Tremarctos ornatus* (Cuvier 1825)), fully herbivorous (giant panda, *Ailuropoda melanoleuca* (Davis 1869)), insectivorous (sloth bear, *Melursus ursinus* (Shaw 1791)), and carnivorous diets (polar bear, *Ursus maritimus* Phipps 1774) (Ewer 1973; Wilson and Mittermeier 2009). Ursidae has an extensive fossil record, starting in the late Eocene of Europe and NA and with a continuous record to the present. In Asia, bears have been recorded from the Oligocene (Hunt 1996; Flynn and Wesley-Hunt 2005), while the family invaded Africa at least three times, the early Miocene (hemicyonines), the late Miocene–early Pliocene (*Indarctos* Pilgrim 1913, *Agriotherium* Wagner 1837), and the Pleistocene (*Ursus* Linnaeus 1758). Ursidae reached SA at least twice during the Pleistocene-Holocene (Hunt 1996; Flynn and Wesley-Hunt 2005; Prevosti and Soibelzon 2012).

Late Eocene–Oligocene ursids are included in a paraphyletic stem group (amphicynodonts) of smaller bears (<15 kg). Hemicyoninae was a group of bears from the Oligocene–Miocene, mainly restricted to northern continents, with long limbs and a digitigrade stance that indicates cursorial habits. Hemicyoninae had a body size range of 5 kg–500 kg and carnassials that retain the shearing function, indicating a more carnivorous diet than modern bears (Hunt 1996). The panda clade is known since the middle Miocene, and the giant panda (*Ailuropoda melanoleuca*) is specialized for a bamboo diet and has a highly modified dentition and skull to suit (Jin et al. 2007; Abella et al. 2012). South American bears belong to the short-faced bear clade (Tremarctinae), recorded since the late Miocene in NA (Hunt 1996; Soibelzon 2004) and since the Pleistocene in the SA.

Arctotherium Burmeister 1879
(Fig. 4.8, Tables 4.1 and 4.2)

Species in South America: *Arctotherium angustidens* Gervais and Ameghino 1880, *A. bonariense* (Gervais 1852), *A. tarijense* Ameghino 1902, *A. vetustum* Ameghino 1885 and *A. wingei* Ameghino 1902 (Soibelzon 2004).

Temporal and geographic distribution: *Arctotherium* is recorded from the Ensenadan to the end of the Pleistocene (Soibelzon and Schubert 2011; Prevosti and Martin 2013). *Arctotherium angustidens* is restricted to the Ensenadan of the Pampean Region, Argentina. Finds of this species in Bolivia lack precise stratigraphic data, as do the records of *A. wingei* and *A. tarijense* (Soibelzon and Schubert 2011). *Arctotherium tarijense* was mentioned in Lujanian beds of southern Patagonia in Chile, Uruguay, and the Pampean Region in Argentina (Soibelzon and Schubert 2011), but the remains from southern Chile are rather fragmentary for a specific determination (Prevosti et al. 2003; Prevosti and Martin 2013; see also López Mendoza et al. 2015). *Arctotherium wingei* has records for the Lujanian of Venezuela and the "Lujanian" of Brazil, and *A. bonariense* for the Lujanian of the Pampean Region in Argentina (Soibelzon and Schubert 2011; Rodrigues et al. 2014). The presence of *A. tarijense, A. bonariense*, and *A. vetustum* in the Bonaerian of the Pampean Region is highly questionable and is based on specimens with poor stratigraphic data. A Lujanian Age cannot be excluded for these fossils.

Fig. 4.8 Lateral view of the
skull and mandible of
Arctotherium vetustum (MMP
1233 M) (**a**) and life
reconstruction of
Arctotherium (**b**).
Scale = 5 cm

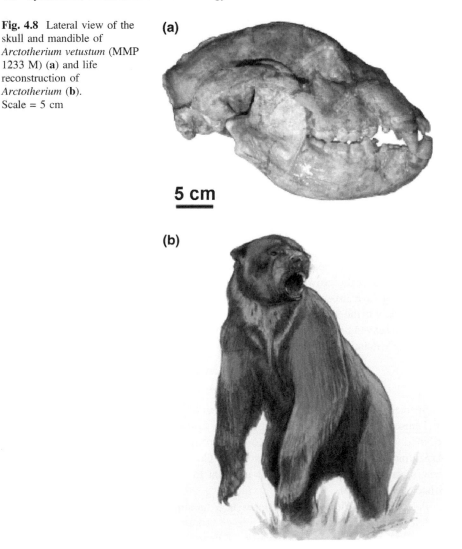

Paleoecology: The largest bear in SA is *Arctotherium angustidens,* with a body
mass of about 580 kg–1750 kg (Soibelzon and Schubert 2011). This is followed by *A.
wingei*: 42 kg–107 kg; *A. tarijense*: 102 kg–224 kg; *A. bonariense*: 100 kg–545 kg;
and *A. vetustum*: 95.5 kg–145 kg (Prevosti and Vizcaíno 2006; Soibelzon and
Tarantini 2009). Morphology, pathology, dental wear pattern, and stable isotopes
suggest that the large *A. angustidens* was omnivorous, eating also carcasses and
bones of large mammals (Soibelzon and Schubert 2011). Other species are interpreted
as omnivorous with a larger proportion of vegetable matter in their diets than the
Ensenadan species (Figueirido and Soibelzon 2009). A paleoecological study based
on southern Patagonian specimens revealed a diet with a large proportion of meat,

probably as carrion from large mammals (Prevosti and Martin 2013). Figueirido and Soibelzon (2009), Soibelzon and Schubert (2011) and Soibelzon et al. (2014) correlated smaller sizes with a tendency to a less carnivorous diet with the increase in abundance of large hypercarnivores after the Ensenadan. Unfortunately, the chronological succession of the species of *Arctotherium* is not corroborated by cladistic analysis (i.e., there is no evidence that they represent an evolutionary sequence), and the supposed lower diversity of large hypercarnivores during the Ensenadan is not supported by the available data (see Prevosti and Soibelzon 2012; Prevosti and Pereira 2014).

Comments: *Arctotherium* is endemic to South America and its presence could be explained by a single immigration event from Central America, in the early–Middle Pleistocene (Soibelzon and Prevosti 2007; Prevosti and Soibelzon 2012). A supposed deciduous lower carnassial (dp4) from the late Pliocene of El Salvador assigned to cf. *Arctotherium* was used to support an origin of *Arctotherium* outside South America (Soibelzon et al. 2008). However, the specimen was not compared with other large Neogene bears from NA (i.e., *Agriotherium* and *Indarctos*) and some features of the crown (e.g., shape, thick enamel with furrows) indicate that it may belong to a juvenile mastodont. A recent ancient DNA study indicates that *Arctotherium* is the sister taxa of *Tremarctos*, instead of *Arctodus*, and suggests that the large scavenger morphotype originated independently in North and South America (Mitchell et al. 2016).

Tremarctos Gervais 1855
(Tables 4.1 and 4.2)

Species in South America: *Tremarctos ornatus* (Cuvier 1825).

Temporal and geographic distribution: *Tremarctos ornatus* is currently distributed along the Andean rain forests, from western Venezuela to northwest Argentina (Wilson and Mittermeier 2009; Cosse et al. 2014). Its fossil record is very scarce and includes Holocene material from Peru and Ecuador (Stahl 2003; Stucchi et al. 2009).

Paleoecology: *Tremarctos ornatus* is a medium-sized bear of 60 kg–200 kg, with an omnivorous diet, including a large proportion of vegetables and occasional large mammals obtained by predation (Wilson and Mittermeier 2009).

Comments: The record of *T. ornatus* in SA during the Holocene represents a second wave of bear migration. The fossil record in NA is much more extensive than in SA and goes back to the late Pliocene (Hunt 1996, 1998).

Mustelidae Fischer 1817

Mustelids are small carnivores with elongated bodies, short limbs, and a broad spectrum of diets (omnivores to hypercarnivores). Locomotion (e.g., scansorial, semifossorial, and aquatic) is correlated with their diversity in skeletal shape (Ewer 1973; Wilson and Mittermeier 2009). Today, the group has a near worldwide distribution. They are first recorded in the late Eocene of Europe and North America; the early Miocene produced the oldest records of the living subfamilies

(Wolsan 1993; Hunt 1996; Baskin 1998). South American mustelids represent four subfamilies that originated in northern continents and entered SA independently (Eizirik 2012; Prevosti and Soibelzon 2012). The subfamilies recorded in South America are the aquatic Lutrinae (otters), Guloninae (e.g., wolverine and martens), Mustelinae (weasels), and Ictonychinae (grisons) (see Eizirik 2012; Sato et al. 2012). A supposed early mustelid found in the late Miocene of La Pampa, Argentina (Verzi and Montalvo 2008) was re-identified as a didelphimorphian marsupial (Prevosti and Pardiñas 2009), while the oldest record of this family in SA is from the Vorohuean Subage (Marplatan, late Pliocene; Berman 1994; Prevosti and Soibelzon 2012).

Mustela Linnaeus 1758
(Tables 4.1 and 4.2)

Species in South America: *Mustela africana* Desmaret 1818, *M. felipei* Izor and de la Torre 1978 and *M. frenata* Lichtenstein 1831.

Temporal and geographic distribution: Currently, *Mustela frenata* is distributed from NA to northwestern SA, *M. felipei* is limited to the Andes of Colombia and Ecuador, and *M. africana* is endemic to the tropical forests of Amazonia (Wilson and Mittermeier 2009). The SA fossil record is very scarce, consisting only of an upper P4 referred to *Mustela* cf. *M. frenata* from the middle Holocene site of La Calera, Ecuador (ca. 4.2 ka BP) and a mention for the middle Holocene archaeological site of Cotocallo, Ecuador (Stahl 2003). *Mustela* was also reported from the late Holocene archaeological site of La Chimba, Ecuador (Stahl 2003).

Paleoecology: Living representatives are small hypercarnivores (0.080 kg–0.450 kg) that eat mainly small rodents (Wilson and Mittermeier 2009).

Comments: *Mustela* migrated into South America at least two times. One event included *M. frenata*, with fossils in NA since the late Pliocene (Anderson 1984). The second event resulted in the entry of the common ancestor of *M. africana* and *M. felipei,* an endemic SA clade. The scarce fossil record of *Mustela* in SA is probably related to a bias of the fossil record against tropical forests (Chap. 5).

Eira Smith 1842
(Tables 4.1 and 4.2)

Species in South America: *Eira barbara* (Linnaeus 1758).

Temporal and geographic distribution: *Eira barbara* currently occurs from Veracruz, Mexico, to northwestern Argentina and southern Brazil and northeastern Argentina to northern Uruguay across tropical and subtropical forests (Wilson and Mittermeier 2009). The fossil record of *Eira* is limited to the "Lujanian" of Brazil, and a dubious mention for Tarija, Bolivia (Lessa et al. 1998; Cartelle 1999; Soibelzon and Prevosti 2007; Prevosti and Soibelzon 2012). A record of *Eira* cf. *E. barbara* from Bahia state in Brazil was dated to between 22–8 ka BP (Castro et al. 2014).

Paleoecology: *Eira barbara* is a medium-sized carnivore (2.7 kg–7 kg) with an omnivorous diet including insects, fruits, and small mammals (Wilson and Mittermeier 2009).

Comments: *Eira* is the only gulonine in SA (Eizirik 2012; Sato et al. 2012), but it is not clear if the genus originated in South America, and later invaded Central America, or if it originated in Central America (Prevosti and Soibelzon 2012).

Galictis Bell 1826

(Fig. 4.9c, Tables 4.1 and 4.2)

Species in South America: *Galictis cuja* (Molina 1782), *G. hennigi* (Rusconi 1931), *G. sorgentinii* Reig 1958, and *G. vittata* (Schreber 1776).

Temporal and geographic distribution: *Galictis vittata* currently inhabits tropical forests from Mexico to northeastern Argentina and *G. cuja* more open environments from northeast Brazil to southern Patagonia (Wilson and Mittermeier 2009; Bornholdt et al. 2013). The extinct species *G. sorgentinii* and *G. hennigi* are limited to the Vorohuean (late Pliocene) and Ensenadan (early–Middle Pleistocene) of Buenos Aires, Argentina, respectively (Reig 1957; Berman 1994; Soibelzon and Prevosti 2007). *Galictis vittata* was found in the "Lujanian" of Minas Gerais and northern Brazil (Cartelle 1999; Rodrigues et al. 2016). Other fossils similar to *G. vittata*, with a metaconid in the lower carnassial, come from the late Pliocene of Buenos Aires, and one without metaconid in the lower carnassial from the "Ensenadan" of Tarija, Bolivia (Werdelin 1991; Berman 1994; Soibelzon and Prevosti 2007). *G. cuja* was recorded in the Late Pleistocene of Minas Gerais, Brazil, and the Holocene of Argentina (Buenos Aires, Jujuy, Neuquén, and Río Negro provinces) and probably Chile (Ultima Esperanza; Massoia 1992; Prevosti and Pardiñas 2001; Quintana 2001; Prevosti and Soibelzon 2012; Rodriges et al. 2016).

Paleoecology: The genus *Galictis* includes small carnivores (1 kg–3.3 kg) with hypercarnivorous habits, hunting small mammals, mainly rodents; Wilson and Mittermeier (2009).

Comments: The taxonomic validity of the fossil species (e.g., *G. hennigi*) and some informally recognized taxa (see Berman 1994) are pending review. The fossil record suggests that this genus originated in SA and, if so, *G. vittata* secondarily migrated to Central America (Prevosti and Soibelzon 2012).

Lyncodon Gervais 1845

(Tables 4.1 and 4.2)

Species in South America: *Lyncodon bosei* Pascual 1958 and *L. patagonicus* (de Blainville 1842).

Temporal and geographic distribution: *Lyncodon bosei* is only known from the type specimen found in the Ensenadan of Buenos Aires (ca. 1 Ma BP). The living species *L. patagonicus* has records in the Lujanian and Platan (Late Pleistocene–Holocene) of Patagonia and the Pampean Region (Pascual 1958; Prevosti and Pardiñas 2001). Currently, this species is distributed in Patagonia (southern Chile

Fig. 4.9 Lateral view of the skull of *Pteronura brasiliensis* (CICYTTP-PV-M-1-21) (**a**), lateral view of the skull of *Stipanicicia pettorutii* (MACN Pv 14260, Holotype) (**b**), and life reconstruction of *Galictis* (**c**). Scale = 5 cm

and Argentina), central areas of Argentina and along the Andes (Wilson and Mittermeier 2009). The fossil record shows that, in connection with climatic changes during the Late Pleistocene and Holocene glaciations, *L. patagonicus* expanded its range to the northeast, including the Pampean Region (Prevosti and Pardiñas 2001; Schiaffini et al. 2013a).

Paleoecology: *Lyncodon patagonicus* is a small carnivore (0.200 kg–0.250 kg) with poorly known habits, that predates on birds and rodents and is probably also capable of hunting fossorial rodents (Prevosti and Pardiñas 2001; Wilson and Mittermeier 2009). *Lyncodon patagonicus* has derived dental traits in comparison to *L. bosei* (e.g., reduction of P4 lingual shelf, loss of m2 and P2/p2; Pascual 1958), which could be related to more carnivorous habits.

Comments: Based on the distribution of the living species and its fossil record, *Lyncodon* originated in SA.

Stipanicicia Reig 1956
(Fig. 4.9b, Tables 4.1 and 4.2)

Species in South America: *Stipanicicia pettorutii* Reig 1956.

Temporal and geographic distribution: Ensenadan of the Buenos Aires, Argentina (Soibelzon and Prevosti 2007). The presence of this genus in older (Marplatan) times is not corroborated by the available data.

Paleoecology: *Stipanicicia pettorutii* was a small mustelid, similar in size to *Galictis,* with a dentition that suggests a hypercarnivorous diet. Based on muscle reconstructions, Ercoli (2015, 2017) suggested that *S. pettorutii* was able to hunt rodents larger than its own body size using a roll and curl strategy to access burrows like *Mustela nigripes.*

Comments: Berman (1994) assigned a lower jaw from southern Buenos Aires province to *Stipanicicia*; however, only cranial material is available for comparison, and unfortunately the mandible has been lost. *Stipanicicia* shares some similarities with *Lyncodon* (e.g., reduction of M1, narrow postorbital constriction; absence of P2) that could indicate a close phylogenetic relationship, something which should be tested with a cladistic analysis. The fossil record suggests that this genus originated in South America.

Lontra Gray 1843
(Tables 4.1 and 4.2)

Species in South America: *Lontra felina* (Molina 1782), *L. longicaudis* (Olfers 1818), and *L. provocax* (Thomas 1908).

Temporal and geographic distribution: *Lontra longicaudis* occurs from Mexico to Uruguay and northeastern Buenos Aires in Argentina; *Lontra felina* lives in the Pacific and the southern parts of the Atlantic coasts of Patagonia in Chile and Argentina, and *L. provocax* inhabits lakes and rivers of Patagonia in Argentina, and the Beagle Channel (Wilson and Mittermeier 2009). *Lontra longicaudis* is known from the Ensenadan–Holocene of Buenos Aires, Argentina (ca. 1 Ma BP), the Lujanian of Uruguay (27–58 ka BP) and the "Lujanian" of Bahia and Minas Gerais, Brazil (Cione and Tonni 1978; Berman 1994; Soibelzon and Prevosti 2007; Prevosti and Ferrero 2008; Acosta et al. 2015). The presence of this species in the Lujanian of the Buenos Aires (Ameghino 1889) has not yet corroborated. *Lontra provocax* was identified from the Holocene of the Beagle Channel, and one archeological site in Neuquén, Argentina (Massoia 1992; Ercoli 2015). Lindsey and

Seymour (2015) mentioned the presence of *"Lutra"* in a Late Pleistocene site in Ecuador.

Paleoecology: The SA species are medium-sized mustelids (3.2 kg–15 kg), with *L. felina* the smallest and *L. provocax* the largest, eating mainly fish, crustaceans, and molluscs (Wilson and Mittermeier 2009).

Comments: Considering the geographic distribution of *L. felina, L. provocax,* and *L. longicaudis,* and that these species form a monophyletic clade, it is possible that *Lontra* reached SA once and *L. longicaudis* later migrated to Central America, or alternatively that the group entered SA twice, as represented by *L. longicaudis* and the common ancestor of *L. felina* and *L. provocax* (Eizirik 2012 for the phylogenetic analysis). The origin of this clade at ca. 3 Ma (see Eizirik 2012) agrees with the first hypothesis and with the conventional time of formation of the Panama Bridge.

Pteronura Gray 1837
(Fig. 4.9a; Tables 4.1 and 4.2)

Species in South America: *Pteronura brasiliensis* (Gmelin 1837).

Temporal and geographic distribution: *Pteronura brasiliensis* inhabits lowland tropical basins of SA, up to northern Uruguay and Entre Ríos in Argentina (Wilson and Mittermeier 2009). Fossils are limited to the last interglacial (ca. 125 ka BP) of Entre Ríos, and the "Lujanian" of Bahia and Mato Grosso in Brazil (Prevosti and Ferrero 2008).

Paleoecology: *Pteronura brasiliensis* is a large mustelid (22 kg–32 kg) that preferentially eats fishes (Wilson and Mittermeier 2009).

Comments: *Pteronura* is the sister taxon of *Satherium* Gazin 1934, an otter from the Pliocene–early Pleistocene of USA. *Pteronura* represents another lineage of otters that invaded the continent from NA (Prevosti and Ferrero 2008).

Mephitidae Bonaparte 1845

Molecular studies indicate that mephitids are distinct from mustelids (Eizirik 2012). Mephitids are small carnivores with an omnivorous diet, aposematic skin coloration, well-developed odor glands, and terrestrial-semifossorial habits (Ewer 1973; Wilson and Mittermeier 2009). Currently they are restricted to the Americas, with the exception of the Asiatic stink badger (*Mydaus* Cuvier 1821) that is the sister taxon of other living mephitids (Wilson and Mittermeier 2009; Eizirik et al. 2010). *Conepatus* Gray 1837, originated in North America, with fossils from the early Pliocene (Baskin 1998; Wang and Carranza-Castañeda 2008) and invaded SA probably in the early–Middle Pleistocene (Berman 1994; Prevosti and Soibelzon 2012).

Conepatus Gray 1837
(Fig. 4.10; Tables 4.1 and 4.2)

Species in South America: *Conepatus altiramus* Reig 1952, *Co. chinga* (Molina 1782), *Co. cordubensis* (Ameghino 1889), *Co. mercedensis* Gervais and Ameghino

Fig. 4.10 Lateral view of the skull and mandible of *Conepatus primaevus* (MSP 1) (**a**) and life reconstruction of *Conepatus chinga* (**b**). Scale = 5 cm

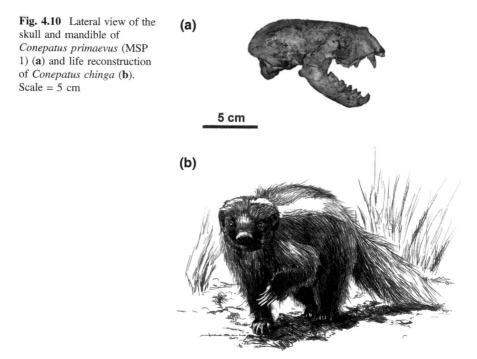

(a)

5 cm

(b)

1880, *Co. primaevus* (Burmeister 1866), *Co. semistriatus* (Boddaert 1785), and *Co. talarae* (Churcher and Van Zyll de Jong 1965).

<u>Temporal and geographic distribution</u>: The two living SA skunks have separate distributions: *Co. chinga* in the southern parts from Peru, Bolivia, Paraguay, and southern Brazil to the Magellan strait in Patagonia and *Co. semistriatus* in the northern part of the continent (Colombia, Venezuela, Ecuador, and North of Peru, and Brazil) and Central America (Wilson and Mittermeier 2009; Schiaffini et al. 2013b). *Conepatus primaevus* is recorded from the Ensenadan and Bonaerian? of Buenos Aires, Argentina (Forasiepi 2003), while *Co. semistriatus* and *Co. talarae* are known from the Lujanian of Muaco, Venezuela and Talara, Peru (13–14 ka. BP; González et al. 2010), respectively. *Conepatus semistriatus* is also found in the "Lujanian" of Bahia, Brazil (Cartelle 1999). *Conepatus mercedensis* occurred in the Ensenadan and, following the stratigraphy of the Luján Basin according to Toledo (2011), the species is also found in the Lujanian of Buenos Aires, Argentina (Berman 1994). *Conepatus chinga* was described from the "Ensenadan" of Tarija, Bolivia, Lujanian–Platan of Buenos Aires, "Lujanian" of Minas Gerais, Brazil, and the late Holocene of Patagonia, Chile, and Argentina (Ameghino 1889; Berman 1994; González et al. 2010). *Conepatus chinga* was also included in the faunal list of the late Holocene archaeological site of La Chimba, Ecuador (Stahl and Athens 2001). *Conepatus cordubensis* is limited to its type locality in the "Bonaerian" of Córdoba, Argentina (Ameghino 1889), but this is probably a junior synonym of *Co. chinga*. Records of *Conepatus* sp. are from the Lujanian of Valdivia, Chile;

Arequipa, Talara, Tirapata, and Huánuco in Peru, and the Ensenadan–Platan of the Buenos Aires and Platan in Patagonia, Argentina (Berman 1994; Tonni et al. 2002; Bonomo 2005; Shockey et al. 2009; Alvarez 2009; González et al. 2010; Lindsey and Seymour 2015). Massoia (1992) and Quintana (2001) mentioned remains of *Co. chinga* from archeological sites of Neuquén and Buenos Aires, respectively. This species probably appears in other archeological sites in Argentina.

The first appearance datum of *Conepatus* in SA is still under debate. *Conepatus altiramus* was described by Reig (1952) and originally suggested that the specimen was recovered from levels 3–5 of the Chapadmalal "Formation" (ca. 3.3 Ma BP), at Barranca de los Lobos, 700 m north of Baliza Caniú (38° 7′ S, 57° 6′ W; Buenos Aires, Argentina) (see Wang and Carranza-Castañeda 2008 who recently defended this age). In 1957, after conversation with Galileo J. Scaglia, who collected the fossil in 1939, Reig suggested that *Co. altiramus* was likely coming from the overlying Barranca de los Lobos "Formation," in view of the fact that this unit is better exposed in the area than the Chapadmalal "Formation." The presence of Pleistocene sediments (Isla et al. 2015) was discounted. In 1958, Reig visited the locality with Galileo J. Scaglia, and re-considered that the holotype of *C. altiramus* effectively was collected from the Chapadmalal "Formation." Additionally, museum label on the type specimen (MMP 173 S) was clearly made by the time of or after the description of *C. altiramus* in 1952, and indicates that it comes from 200 m south of the Punta Mala ("*Costa Atlántica, 200 metros al S. de la bajada de Punta Mala*") from the base of the cliff of the Chapadmalal "Formation." This information is similar to that reported by Reig (1952) and agrees with the information on other fossil labels (Olivares et al. 2012; U.F.J. Pardiñas, pers. comm.), suggesting that there was a change in the geographic point of reference (Baliza Caniú vs. Punta Mala). Considering the likely congruence between the data presented by Reig (1952, 1958) and the MMP label, and evaluating the geological data (e.g., Isla et al. 2015) only the Chapadmalal "Formation" is exposed in the lower section of the locality where *C. altiramus* comes from (but Pleistocene deposits are present in the upper section). Based on the unclear situation about the stratigraphic provenance and the absence of new skunk material from the Chapadmalal "Formation" (or other deposits older than the Ensenadan), several authors considered that the age of this taxon is uncertain and that the confirmed records of skunks in South America are limited to the Ensenadan or younger ages (e.g., Berman 1994; Cione and Tonni 1995; Prevosti and Soibelzon 2012), with whom we agree. The evidence is not clear enough to extend the fossil range of the skunk lineage in SA into the Pliocene (see also Prevosti and Pardiñas, in press) because (1) the fossil was collected in 1939, more than 10 years before the publication of *C. altiramus* and the establishment of the main stratigraphic scheme in the region (Kraglievich 1952); (2) there is no original label in the MMP or field notes written when the fossil was collected to corroborate its provenance; (3) there are no other remains of this taxa or confirmed skunks found in levels older than the Ensenadan. In addition, considering the geographic provenance, it is also possible that the specimen was collected from younger Pleistocene rocks, including a dislocated rocks from upper levels. Wang and Carranza-Castañeda (2008) erroneously indicate that *Co. talarae* is present in the early–Middle Pleistocene of SA, but this taxon is only known from the type locality, which is dated to the Late Pleistocene (see above).

Paleoecology: *Conepatus* group's small carnivores (1–3.5 kg), with an omnivorous diet, are composed of insects, fruits, small vertebrates, and carrion (Wilson and Mittermeier 2009).

Comments: The oldest fossils of *Conepatus* are from the Pliocene of Mexico (4–5 Ma BP) which suggests a NA origin for the clade (Wang and Carranza-Castañeda 2008). The area of origin of *Co. semistriatus* is uncertain (Central or South America); other SA species are clearly from the southern continent by reference to fossil record (Wang and Carranza-Castañeda 2008; Prevosti and Soibelzon 2012). The presence of *Co. primaevus* in the Ensenadan is based on the interpretation of Kraglievich (1934), but at present it is difficult to corroborate this interpretation. On the other hand, the presence of *Co. mercedensis* in the Ensenadan is based on the inclusion of *Conepatus mercedensis praecursor* Rusconi 1932 as a subspecies of this species (Berman 1994).

Procyonidae Bonaparte 1850

Procyonids are, with few exceptions, small- to medium-sized carnivores that inhabit mostly forests, or semiforested habitats, with omnivorous diets and a generalized postcranial structure with scansorial capabilities (Ewers 1973; Wilson and Mittermeier 2009). They are restricted to tropical and subtropical regions of the Americas and are presently absent from other continents (Wilson and Mittermeier 2009). The fossil record shows that their earliest remains are from the Oligocene of Europe, and if Simocyoninae are in fact part of this family, then it was present in other Holarctic continents in the Miocene (Hunt 1996). Miocene–Pliocene NA procyonids were very diverse, including the groups that have invaded South America since the late Miocene (Baskin 1998, 2003). Rodriguez et al. (2013) interpreted that procyonids immigrated twice to South America (in the late Miocene to give rise to *Cyonasua* Ameghino 1885 and *Chapalmalania* Ameghino 1908, and again in the Late Pleistocene for the living taxa), which is inconsistent with phylogenetic reconstructions that indicate multiple immigration events (Prevosti and Soibelzon 2012). The oldest records of *Potos* E. Geoffroy Saint-Hilaire and Cuvier 1795, and *Nasuella* Hollister 1915, are part of the faunal list of the late Holocene archaeological site of La Chimba, Ecuador (Stahl 2003). *Bassaricyon* Allen 1876, has no fossil record.

Nasua Storr 1870
(Tables 4.1 and 4.2)

Species in South America: *Nasua narica* (Linnaeus 1766) and *Nasua nasua* (Linnaeus 1766).

Temporal and geographic distribution: *Nasua narica* currently inhabits Central America and northwest Colombia, while *N. nasua* is distributed from Colombia and Venezuela in the north, to northern Uruguay and northern Argentina in the south (Wilson and Mittermeier 2009). Fossil remains of *N. nasua* has have been found in the "Lujanian" of Minas Gerais and Bahia and Late Pleistocene of Tocantins (Rodrigues et al. 2014), Brazil, and a putative record of *Nasua* from the "Ensenadan" of Tarija, Bolivia (Soibelzon and Prevosti 2007, 2013). *Nasua* cf. *N. nasua* was included in the faunal list of the late Holocene archeological site of La Chimba, Ecuador (Stahl 2003) and described from a late Holocene (ca. 500 years BP) archeological site in northeastern Buenos Aires, Argentina. This record is outside the current geographic distribution of the genus, and could be explained by either anthropic transportation or by a wider distribution of the taxon in the past, associated with recent climatic changes (Ramírez et al. 2015).

Paleoecology: *Nasua* includes small carnivores (2 kg–7 kg) with omnivorous diets, composed mainly of invertebrates and fruit, and with good scansorial capabilities (Wilson and Mittermeier 2009).

Comments: The fossil record of *Nasua*, species distribution, and phylogenetic reconstructions suggest that the genus originated outside SA, with the exception of *N. nasua* (Prevosti and Soibelzon 2012). Molecular data suggest that *Nasuella* is a junior synonym of *Nasua* (Eizirik 2012).

Procyon Storr 1870

(Tables 4.1 and 4.2)

Species in South America: *Procyon cancrivorus* (Cuvier 1798), *Procyon lotor* (Linnaeus 1758).

Temporal and geographic distribution: *Procyon cancrivorus* occurs from Panama to Uruguay and northeastern Buenos Aires (Argentina), while *P. lotor* in SA is only recorded in the Caribbean region of Colombia (Wilson and Mittermeier 2009; Fracassi et al. 2010). Fossil raccoons have been found in the Late Pleistocene of Bahia, Minas Gerais and Tocantins, Brazil and the Lujanian of Formosa, northern Argentina (Soibelzon and Prevosti 2007, 2013; Rodriguez et al. 2013). The latter is associated with a Late Pleistocene fauna, but the correlation made by Soibelzon et al. (2010) with other locality dated ca. 60 ka has no support (see discussion in Prevosti and Schubert 2013). *Procyon* cf. *P. cancrivorus* was listed for the late Holocene La Chimba archaeological site of Ecuador (Stahl 2003), while unpublished *Procyon* remains have been recovered from the Middle Pleistocene Orocual archaeological site in eastern Venezuela (Prevosti *pers. obs.*). As is discussed elsewhere (Soibelzon and Prevosti 2007), the absence of a stratigraphic context or dates associated with the Brazilian fossils does not allow us to confirm their Lujanian Age.

Paleoecology: *Procyon* includes small- to medium-sized carnivores (3.1 kg–7.7 kg), with omnivorous diets (invertebrates, fruits, small vertebrates) and scansorial habits (Wilson and Mittermeier 2009). Currently, *Procyon* inhabits different kind of forested habitats, but also was observed in the South American Llanos.

Comments: The oldest record of *Procyon* is from the late Miocene–Pliocene of NA (Baskin 1998). Soibelzon (2011) and Rodriguez et al. (2013) inferred that raccoons migrated to SA in the Late Pleistocene; however, this could be an artifact of the bias against tropical areas in the fossil record, especially before the Late

(a)

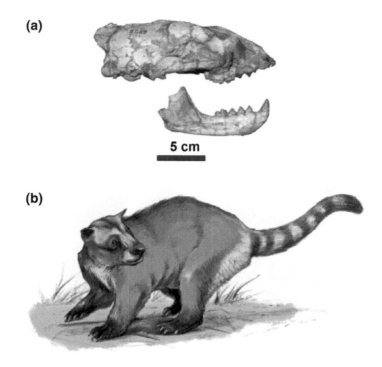

(b)

Fig. 4.11 Lateral view of the skull of *Cyonasua brevirostris* (MACN Pv 8209, Holotype of *Amphinasua longirostris*), and life reconstruction of *Cyonasua* (**b**). Scale = 5 cm

Pleistocene (Prevosti and Soibelzon 2012; Prevosti et al. 2013), as suggested by the unpublished *Procyon* specimens from the Middle Pleistocene of Venezuela. *Procyon cancrivorus* may have its origin in South America (Prevosti and Soibelzon 2012).

Cyonasua Ameghino 1885
(Fig. 4.11; Tables 4.1 and 4.2)

Species in South America: *Cyonasua argentina* Ameghino 1885, *Cyonasua brevirostris* (Moreno and Mercerat 1891), *Cyonasua clausa* (Ameghino 1904), *Cyonasua groeberi* JL Kraglievich and Reig 1955, *Cyonasua lutaria* (Cabrera 1936), *Cyonasua meranii* (C. Ameghino and Kraglievich 1925), and *Cyonasua pascuali* Linares 1982.

Temporal and geographic distribution: The fossil record of *Cyonasua* principally includes Argentina, and scarce material from Bolivia, and Venezuela from the late Miocene (Huayquerian) to the Middle Pleistocene (Ensenadan) (Soibelzon and Prevosti 2007, 2013; Prevosti and Soibelzon 2012; Forasiepi et al. 2014). The oldest specimen of this genus (*Cyonasua* sp.) is from Catamarca province, Argentina, and has an age of between 8.7 and 7.14 Ma (Esteban et al. 2014), while the youngest belongs to *Cy. meranii* and was collected in the Ensenadan of Buenos

Aires Province (ca. 1 Ma; Soibelzon and Prevosti 2007, 2013; Prevosti and Soibelzon 2012). Several species have been recorded in the Huayquerian: *Cy. argentina* in Entre Ríos and possibly Catamarca, *Cy. pascuali* in Mendoza, and *Cy. brevirostris* in Catamarca and La Pampa (Berta and Marshall 1978; Berman 1989). *Cyonasua lutaria* is present in the Chapadmalalan (ca. 3.3 Ma) of Buenos Aires and possibly in the Pliocene of Catamarca (Kraglievich and Reig 1954; Kraglievich and Olazábal 1959; Berta and Marshall 1978; Marshall et al. 1979; Berman 1989). *Cyonasua clausa* is only known from the Montehermosan (early Pliocene) of Buenos Aires, and *Cy. groeberi* from the late Miocene or Pliocene of Córdoba (Kraglievich and Reig 1954; Berta and Marshall 1978). *Cyonasua* is also recorded in the Huayquerian of Jujuy, in the late Pliocene of Venezuela, and in the late Miocene of Bolivia (Berman 1989; Forasiepi et al. 2014; Poiré et al. 2015). Procyonid postcranial remains that could belong to *Cyonasua* have been found in the late Miocene–early Pliocene of Peru (Soibelzon and Prevosti 2013).

Paleoecology: The estimated body mass of *Cyonasua* using the lower carnassial is about 6 kg (Prevosti et al. 2013); however, this value seems to be an underestimate because this taxon has a small lower m1 and skull measurements indicates a larger body size. We made new estimation using the condylobasal length of the skull (LCB), and the distance between the skull condyles and the anterior border of the orbits (LOO), using the formulae of Van Valkenbrugh (1990) based on a sample of living Carnivora. We obtained body mass estimates of 20.5 kg (SKL) and 23.4 kg (OOL) for the holotype of *Cyonasua brevirostris* (MLP 10-52), and 21.8 kg (OOL) for an incomplete skull of *Cy. lutaria* (MMP S 369). Using geometric similarity and the mean of these estimations, we inferred the body size of other species: *Cyonasua argentina*: 19.6 kg; *Cyonasua pascuali*: 17.04 kg, *Cyonasua groeberi*: 23.1 kg, *Cyonasua clausa*: 23.9 kg, *Cyonasua lutaria*: 21.8 kg; *Cyonasua meranii*: 20.6 kg. The estimates for some species (e.g., *Cy. lutaria* and *Cy. brevirostris*) accord better with the body masses of some living species (e.g., *Gulo gulo* (Linnaeus 1758) and *Neofelis nebulosa* (Griffith 1821), with a body mass of ca. 20 kg and 22 kg, respectively) that have similar cranial measurements, but not with others (e.g., *Mellivora capensis* (Schreber 1776)) and *Canis latrans* Say 1823, with a body mass of ca. 10 kg and 13 kg, respectively; see Van Valkenburgh 1990). The results obtained with geometric similarity are likely overestimates, because they are larger than the body mass of *Procyon lotor* (ca. 8 kg–10 kg) despite its having similar dental measurements. Thus, we consider that a mean between the dental and cranial estimators is the best way to reduce bias (Table 4.2). Using this procedure, the body mass of *Cyonasua* is between 15 kg (*Cy. clausa*) and 11 kg (*Cy. pascuali*). Using postcranial measurements of some species of *Cyonasua*, preliminary work by Tarquini et al. (2015; 2016) obtained body masses between 13 kg–25 kg, which overlap with the ones obtained here using cranial measurements and the "mean approach". Wroe et al. (2004) reported a body mass of 23.7 kg for *Cy. argentina* based on the Van Valkenburgh (1990) SKL equation. Unfortunately, there is no *Cyonasua* skull confidently assigned to *Cy. argentina*, and we cannot identify which specimen was used in this analysis. Dental morphology suggests a hypocarnivorous and omnivorous diet (Prevosti et al. 2013;

Soibelzon and Prevosti 2013). Soibelzon (2011) inferred that *Cyonasua* had a more predatory niche than living South American procyonids. However, shearing structures useful to process meat are reduced in the dentition of *Cyonasua* suggesting an omnivorous and generalist diet. A recent study of postcranial elements of *Cyonasua* (Tarquini et al. 2017) suggests that this taxon had generalized locomotor habits, with some degree of grasping ability, congruent with climbing capabilities.

Comments: The fossil record and phylogenetic position suggest that *Cyonasua* evolved in South America (Baskin 1989, 2004; Prevosti and Soibelzon 2012; Forasiepi et al. 2014). Recently, several rodent remains (i.e., *Actenomys* sp., *Microcavia* sp.,) and small notoungulates (i.e., *Paedotherium*) have been found in a paleocave, in the Chapadmalal "Formation" ("middle" Pliocene) near Miramar (Buenos Aires, Argentina). This association has been interpreted as the product of the depredation of *Cyonasua lutaria* or a carnivorous didelphimorphian (i.e., *Thylophorops chapadmalensis*) (Cenizo et al. 2016) and could indicate that *Cyonasua* preyed on these small mammals.

Chapalmalania Ameghino 1909
(Fig. 4.12; Tables 4.1 and 4.2)

Species in South America: *Chapalmalania altaefrontis* Kraglievich and Olazabal 1959 and *Ch. ortognatha* Ameghino 1908.

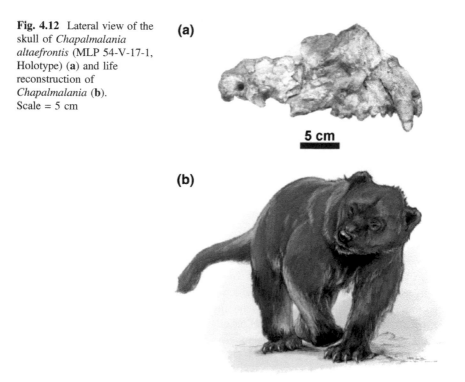

Fig. 4.12 Lateral view of the skull of *Chapalmalania altaefrontis* (MLP 54-V-17-1, Holotype) (**a**) and life reconstruction of *Chapalmalania* (**b**). Scale = 5 cm

(a)

5 cm

(b)

Temporal and geographic distribution: *Chapalmalania altaefrontis* comes from the Pliocene of Mendoza, probably from Chapadmalalan levels, and *Ch. ortognatha* was found in the Chapadmalalan of Buenos Aires (ca. 3.3 Ma; Ameghino 1908; Kraglievich and Olazabal 1959; Berman 1994; Soibelzon and Prevosti 2007, 2013). One specimen of *Chapalmalania* cf. *Ch. altaefrontis* was discovered in the Pliocene of Catamarca and one assigned to *Ch. altaefrontis* in the Vorohuean (late Pliocene) of Buenos Aires (Kraglievich and Olazabal 1959; Marshall et al. 1979; Berman 1994). An isolated upper M1 of *Chapalmalania* sp. was recently found in the late Pliocene of Colombia (Forasiepi et al. 2014).

Paleoecology: The body mass of *Chapalmalania* species was estimated to have been about 22 kg using lower carnassial length (Prevosti et al. 2013). As in the case of *Cyonasua*, this value seems to be an underestimate. However considering the size of the cranial remains and the specimen MMP 1121 M, new estimations using the SKL and OOL equations (Van Valkenburgh 1990), we obtained a body mass of 125 kg–181 kg (mean: 153.76 kg), comparable to small/medium-sized bears (e.g., *Ursus americanus* Pallas 1780). The mean of the dental and cranial estimates (ca. 90 kg), gives a lower value comparable to females of small-sized bears (e.g., *Tremarctos ornatus*; Wilson and Mittermeier 2009). Wroe et al. (2004) reported a body mass of 93.1 kg for *Ch. altaefrontis* based on the Van Valkenburgh (1990) SKL equation, but there is no a complete skull for taking this measurement. An omnivorous scavenger diet was inferred for *Chapalmalania* (Kraglievich and Olazábal 1959; Berman 1994; Soibelzon and Prevosti 2007, 2013). Recently, de los Reyes et al. (2013) interpreted some bite marks found in a caudal vertebra of a Pliocene glyptodont as produced by *Chapalmalania*.

Comments: Species determination is a difficult task. The holotype of *Ch. ortognatha* is an anterior portion of palate with the premaxillary region, incisors, and canines, while the holotype of *Ch. altaefrontis* is a rostrum with complete dentition and a portion of the basicranium. Consequently, the determination of isolated mandibles is subjective. A skull, MMP 1121 M from the Chapadmalalan of Buenos Aires could belong to *Ch. ortognatha* (Prevosti et al. 2013), or to a third species, with the rostrum being clearly longer than in *Ch. altaefrontis*. Following the interpretation for *Cyonasua*, *Chapalmalania* could have originated in SA (Prevosti and Soibelzon 2012).

References

Abella J, Alba DM, Robles JM, Valenciano A, Rotgers C, Carmona R, Montoya P, Morales J (2012) *Kretzoiarctos* gen. nov., the oldest Member of the giant panda clade. 7(11):e48985

Akersten WA (1985) Canine function in *Smilodon* (Mammalia; Felidae; Machairodontinae). Contr Sci Mus Nat Hist Los Angeles 356:1–22

Alvarez MC (2009) Resultados preliminares del análisis arqueofaunístico del sitio Calera (Partido de Olavarría, Región Pampeana). In: Bourlot T, Bozzuto D, Crespo C, Hecht AC, Kuperszmit N (eds) Entre pasados y presentes II : studios contemporáneos en ciencias antropológicas. Fundación de Historia Natural Félix de Azara, Buenos Aires, pp 307–322

Ameghino F (1889) Contribución al conocimiento de los mamíferos fósiles de la República Argentina. Acad Nac Cien Córdoba, Actas 6:1–1027

Ameghino F (1908) Las formaciones sedimentarias en la región litoral de Mar del Plata y Chapadmalal. An Mus Nac 3:343–428

Amorosi T, Prevosti FJ (2008) A Preliminary review of the Canid remains from Junius Bird's excavations at Fell's and Pali Aike. Curr Res Pleistocene 25:25–27

Anderson E (1984) Review of the small carnivores of North America during teh last 3.5 million years. In: Genoways HH, Dawson MR (eds) Contributions in Quaternary vertebrate paleontology: a volume in memorial to John E. Guilday. Carnegie Museum of Natural History, Pittsburgh, pp 257–266

Andersson K (2004) Elbow-joint morphology as a guide to forearm function and foraging behaviour in mammalian carnivores. Zool J Linn Soc 142:91–104

Andersson K, Norman D, Werdelin L (2011) Sabretoothed carnivores and the killing of large prey. PLoS ONE 6(10):e24971

Antón M, Galobart A (1999) Neck function and predatory behavior in the scimitar toothed cat *Homotherium latidens* (Owen). J Vert Paleontol 19:771–784

Austin JJ, Soubrier J, Prevosti PJ, Prates L, Trejo V, Mena F, Cooper A (2013) The origins of the enigmatic Falkland Islands wolf. Nat Commun 4:1552

Baskin JA (1989) Comments on new world tertiary Procyonidae (Mammalia, Carnivora). J Vert Paleontol 9:110–117

Baskin JA (1998) Mustelidae. In: Janis CM, Scott KM, Jacobs LL (eds) Evolution of Tertiary mammals of North America, volume 1: terrestrial carnivores, ungulates, and ungulatelike mammals, vol 1. Cambridge University Press, Cambridge, pp 152–173

Baskin JA (2003) New procyonines from the Hemingfordian and Barstovian of the Gulf Coast and Nevada, including the first fossil record of the Potosini. Bull Amer Mus Nat Hist 279:125–146

Baskin JA (2004) Bassariscus and Probassariscus (Mammalia, Carnivora, Procyonidae) from the early Barstovian (middle Miocene). J Vert Paleontol 24:709–720

Berman WD (1989) Cánidos cuaternarios de la provincia de Buenos Aires. El registro de *Protocyon* y su distribución temporal. VI Jornadas Argentinas de Paleontología Vertebrados, Actas 1:77–79

Berman WD (1994) Los carnívoros continentales (Mammalia, Carnivora) del Cenozoico en la provincia de Buenos Aires. Unpublished PhD Thesis, Universidad Nacional de La Plata, La Plata

Berta A (1984) The Pleistocene bus dog Speothos pacivorus (Canidae) from the Lagoa Santa caves, Brazil. J Mammal 65:549–559

Berta A (1985) The status of *Smilodon* in North and SouthAmerica. Contr Sci Nat Hist Mus Los Angeles 370:219–227

Berta A (1987) Origin, diversification, and zoogeography of the South American Canidae. Fieldiana Zoology (n.s.) 39:455–471

Berta A (1989) Quaternary evolution and biogeography of the Large South American Canidae (Mammalia: Carnivora). Univ Calif Pub Geol 132:1–149

Berta A (1995) Fossil carnivores from the Leisey Shell Pits, Hillsborough County, Florida. Bull Fla Mus Nat Hist 37:463–499

Berta A, Marshall LG (1978) South American Carnivora. In: Westphal F (ed) Fossilium catalogus, I: Animalia (125). The Hague (Dr. W. Junk), Boston/London, pp 1–48

Biknevicius AR, Van Valkenburgh B (1996) Design for killing: craniodental adaptations of predators. In: Gittleman JL (ed) Carnivore Behavior, Ecology, and Evolution, vol 2. Cornell University Press, New York, pp 393–428

Bocherens H, Cotte M, Bonini R, Scian D, Straccia P, Soibelzon L, Prevosti FJ (2016) Paleobiology of sabretooth cat Smilodon populator in the Pampean Region (Buenos Aires Province, Argentina) around the Last Glacial Maximum: Insights from carbon and nitrogen stable isotopes in bone collagen. Palaeogeogr Palaeoclimatol Palaeoecol 449:463–474

Bonomo M (2005) Costeando las llanuras. Arqueología del litoral marítimo pampeano. Sociedad Argentina de Antropología, Buenos Aires, pp 1–334

Bornholdt R, Helgen K, Klaus-Peter K, Oliveira L, Lucherini M, Eizirik E (2013) Taxonomic revision of the genus *Galictis* (Carnivora: Mustelidae): species delimitation, morphological diagnosis, and refined mapping of geographical distribution. Zool J Linn Soc 167:449–472

Brain CK (1981) The hunters or the hunted? An introduction to African Cave Taphonomy. University of Chicago Press, Chicago, p 376

Brown JG (2014) Jaw Function in Smilodon fatalis: A Reevaluation of the Canine Shear-Bite and a Proposal for a New Forelimb-Powered Class 1 Lever Model. PLoS ONE 9(10):e107456

Cabrera AL (1934) Los perros domésticos de los indígenas del territorio argentino. Actas y trabajos del XXV Congreso Internacional de Americanistas, vol 1, Universidad Nacional de La Plata, La Plata, pp 81–93

Cabrera A, Yepes J (1940) Mamíferos Sud Americanos: Vida, costumbres y descripción. Compañía Argentina de Editores, Buenos Aires

Carbone C, Maddox T, Funston PJ, Mills MGL, Grether GF, Vam Valkenburgh B (2009) Parallels between playbacks and Pleistocene tar seeps suggest sociality in an extinct sabretooth cat, *Smilodon*. Biol Lett 5:81–85

Cardich A (1979) A propósito de un motivo sobresaliente en las pinturas rupestres de "El Ceibo" (Provincia de Santa Cruz, Argentina). Relaciones de la Sociedad de Antropología (n.s.) 13:163–182

Cartelle C (1999) Pleistocene Mammals of the Cerrado and Caatinga of Brazil. In: Eisemberg JF, Redford KH (eds) Mammals of the Neotropics. The Central Neotropics, vol 3. University of Chicago Press, Chicago and London, pp 27–46

Cartelle C, Langguth A (1999) *Protocyon troglodytes* (Lund): Um Canídeo Intertropical Extinto. An Acad Bras Cienc 71:371–384

Castellanos A (1958) Nota preliminar sobre nuevos restos de mamíferos fósiles en el Brocherense del Valle de Los Reartes (provincia de Córdoba, Argentina). Congreso Geológico Internacional (1956) México. Actas 1:217–233

Castro MC, Montefeltro FC, Langer MC (2014) The Quaternary vertebrate fauna of the limestone cave Gruta do Ioiô, northeastern Brazil. Quat Int 352:164–175

Caviglia S (1986) Nuevos restos de cánidos tempranos en sitios arqueológicos de fuego-patagonia. Anales del Instituto de la Patagonoia, Serie Cs, Sociales 16:85–93

Cenizo M, Soibelzon E, Magnussen Saffer M (2016) Mammalian predator–prey relationships and reoccupation of burrows in the Pliocene of the Pampas Region (Argentina): new ichnological and taphonomic evidence. Hist Biol 28:1026–1040

Chimento NR (2016) Sistemática y relaciones filogenéticas de la familia Felidae en América del Sur. Contribuciones del MACN 6:373–383

Chimento NR, Rosa M, Hemmer H (2014) *Puma (Herpailurus) pumoides* (Castellanos, 1958) nov. comb. Comentarios sistemáticos y registro fósil. Estudios Geológicos, Serie Correlación Geológica 30:92–134

Christiansen P (2008) Evolution of skull and mandible shape in cats (Carnivora: Felidae). PLoS ONE 3(7):e2807

Christiansen P (2013) Phylogeny of the sabertoothed felids (Carnivora: Felidae: Machairodontinae). Cladistics 29:543–559

Christiansen P, Harris JM (2005) Body size of *Smilodon* (Mammalia: Felidae). J Morphol 266:369–384

Cione AL, Tonni EP (1978) Paleoethnozoological context of a site of Las Lechiguanas islands, Parana Delta, Argentina. El Dorado. Newsl Bull S Am Anthropol 3(1):76–86

Cione AL, Tonni EP (1995) Chronostratigraphy and "Land mammal-ages": The Uquian problem. J Pal 69:135–159

Clutton Brock J (1996a) Competitors, companions, status symbols or pests: a review of human associations with other carnivores. In: Gittleman JL (ed) Carnivore behavior, ecology, and evolution, vol 2. Cornell University Press, New York, pp 375–392

Clutton Brock J (1996b) The carnivore remains from the excavation at Fell's cave, Chile. In: Hyslop J (ed) Travels and archaeology in South Chile, by JB Bird, University of Iowa Press, Iowa City, pp 188–195

Coltrain JB, Harris JM, Cerling TE, Ehleringer JR, Dearing MD, Ward J, Allen J (2004) Rancho La Brea stable isotope biogeochemistry and its implications for the palaeoecology of late Pleistocene, coastal southern California. Palaeogeogr Palaeoclimatol Palaeoecol 205:199–219

Cosse M, Del Moral Sachetti JF, Mannise N, Acosta M (2014) Genetic evidence confirms presence of Andean bears in Argentina. Ursus 25:163–171

Culver M, Johnson WE, Pecon-Plattery J, O'brien SJ (2000) Genomic ancestry of the American puma. Genetics 91:186–197

de los Reyes M, Poiré D, Soibelzon L, Zurita AE, Arrouy MJ (2013) First evidence of scavenging of a glyptodont (Mammalia, Glyptodontidae) from the Pliocene of the Pampean region (Argentina): taphonomic and paleoecological remarks. Palaeontol Electron 16:1–13

del Papa LM (2012) Una aproximación al estudio de los sistemas de subsistencias a través del análisis arqueofaunístico en un sector de la cuenca del Río Dulce y cercanías a la Sierra de Guasayán. Unpublished Doctoral Thesis, Facultad de Ciencias Naturales y Museo, universidad Nacional de La Plata, La Plata, pp 1–559

Dietz JM (1985) Chrysocyon brachyurus. Mamm Species 234:1–4

Eizirik E (2012) A molecular view on the evolutionary history and biogeography of Neotropical carnivores (Mammalia, Carnivora). In: Patterson B, Costa L (eds) Bones, clones and Biomes. The University of Chicago Press, Chicago, pp 123–142

Eizirik E, Murphy WJ, Koepfli KP, Johnson WE, Dragoo JW, Wayne RK, O'Brien SJ (2010) Pattern and timing of diversification of the mammalian order Carnivora inferred from multiple nuclear gene sequences. Mol Phylogenet Evol 56:49–63

Ercoli MD (2015) Morfología del aparato músculo-esqueletario del postcráneo de los mustélidos (Carnivora, Mammalia) fósiles y vivientes de América del Sur: implicancias funcionales en un contexto filogenético. Unpublished PhD Thesis, Universidad Nacional de La Plata, La Plata

Ercoli MD (2017) Morpho-functional analysis of the mastoid region of the extinct South American mustelid †Stipanicicia pettorutii. Earth Env Sci T R So 116:337–349

Esteban G, Nasif N, Georgieff SM (2014) Cronobioestratigrafía del Mioceno tardío—Plioceno temprano, Puerta de Corral Quemado y Villavil, provincia de Catamarca, Argentina. Acta Geol Lillo 26:165–188

Ewer RF (1973) The carnivores. Cornell University Press, New York

Fariña RA, Czerwonogora ADA, Giacomo MDI (2014) Splendid oddness: revisiting the curious trophic relationships of South American Pleistocene mammals and their abundance. An Acad Bras Cienc 86:311–331

Figueirido B, Soibelzon LH (2009) Inferring palaeoecology in extinct tremarctine bears (Carnivora, Ursidae) using geometric morphometrics. Lethaia 43:209–222

Flynn JJ, Wesley-Hunt GD (2005) Carnivora. In: Rose KD, Archibald JD (eds) The Rise of placental mammals. Origins and Relationships of the Major Clades. The Johns Hopkins University Press, Baltimore and London, pp 175–198

Flynn JJ, Finarelli J, Spaulding S (2010) Phylogeny of the Carnivora and Carnivoramorpha, and the use of the fossil record to enhance understanding of evolutionary transformations. In: Goswami A, Friscia A (eds) Carnivoran evolution: New views on Phylogeny, form and function. Cambridge University Press, Cambridge, pp 25–63

Forasiepi AM (2003) Nuevo registro de Conepatus primaevus (Mammalia, Carnivora, Mustelidae) del Pleistoceno de la Provincia de Buenos Aires, Argentina. Rev Mus Arg Cienc Nat "B. Rivadavia" (n.s.) 5:21–29

Forasiepi AM, Martinelli AG, Blanco JL (2007) Bestiario fósil. Mamíferos del Pleistoceno de la Argentina. Editorial Albatros, Buenos Aires, pp 1–192

Forasiepi AM, Soibelzon LH, Gomez CS, Sánchez R, Quiroz LI, Jaramillo C, Sánchez-Villagra MR (2014) Carnivorans at the Great American Biotic Interchange: new discoveries from the northern neotropics. Naturwissenschaften 101:965–974

Fracassi NG, Moreyra PA, Lartigau B, Teta P, Landó R, Pereira JA (2010) Nuevas especies de mamíferos para el Bajo Delta del Paraná y bajíos ribereños adyacentes, Buenos Aires, Argentina. Mastozool Neotrop 17:367–373

González E, Prevosti FJ, Pino M (2010) Primer registro de Mephitidae (Carnivora: Mammalia) para el Pleistoceno de Chile. Magallania 38:239–248

Gordillo I (2010) La imagen del Felino en la América Precolombina. Grupo Abierto Comunicaciones, Beccar

Guérin C, Hugueney M, Mourer Chauviré C, Faure M (1993) Paléoenvironnement Pléistocène dans l'aire archéologique da São Raimundo Nonato (Piaui, Brésil): apport des mammifères et des oiseaux. Documents des Laboratoires de Géologie de Lyon 125:187–202

Hadler Rodríguez P, Prevosti FJ, Ferigolo J, Ribeiro A (2004) Novos materiais de Carnivora para o Pleistoceno do Estado do Rio Grande do Sul, Brasil. Rev Bras Paleontol 7:77–86

Hemmer H, Kahlke R-D, Vekua A (2004) The Old World puma—*Puma pardoides* (Owen, 1846) (Carnivora: Felidae)—in the lower Villafranchian (upper Pliocene) of Kvabebi (East Georgia, Transcaucasia) and its evolutionary and biogeographical signiicance. Neues Jahrb Geol Paläontol 233:197–231

Hemmer H, Kahlke R-D, Vekua A (2010) *Panthera onca georgica* ssp. nov. from the early Pleistocene of Dmanisi (Republic of Georgia) and the phylogeography of jaguars (Mammalia, Carnivora, Felidae). Neues Jahrb Geol Paläontol 257:115–127

Hildebrand M (1952) An analysis of body proportion in the Canidae. Am J Anat 90:217–256

Hildebrand M (1954) Comparative morphology of the body skeleton in recent Canidae. Univ Calif Publ Zool 52:399–496

Hoffstetter R (1952) Les mammifères pléistocènes de la République de l`Equateur. Mém Soc Géol Fr (n.s.), 31(66):1–391

Hoffstetter R (1986) High Andean mammalian faunas during the Plio-Pleistocene. In: Vuilleumier F, Monasterds M (eds) High altitude sub tropical biogeography. Oxford University Press, Oxford, pp 218–245

Hunt RM Jr (1996) Biogeography of the order Carnivora. In: Gittleman JL (ed) Carnivore behavior, ecology, and evolution, vol 2. Cornell University Press, New York, pp 485–541

Hunt RM Jr (1998) Ursidae. In: Janis CM, Scott KM, Jacobs LL (eds) Evolution of tertiary mammals of North America, vol 1: terrestrial carnivores, Ungulates and Ungulatelike Mammals, vol 1. Cambridge University Press, Cambridge, pp 174–195

Isla F, Taglioretti M, Dondas A (2015) Revisión y nuevos aportes sobre la estratigrafía y sedimentología de los acantilados entre Mar de Cobo y Miramar, provincia de Buenos Aires. Revista de la Asociación Geológica Argentina 72:235–250

Jin C, Ciochon RL, Dong W, Hunt RM, Liu J, Jaeger M, Zhu Q (2007) The first skull of the earliest giant panda. Proc Natl Acad Sci USA 104:10932–10937

Johnson WE, Eizirik E, Pecon-Slattery J, Murphy WJ, Antunes A, Teeling E, O'Brien SJ (2006) The late Miocene radiation of modern Felidae: a genetic assessment. Science 311(5757):73–77

Kiffner C (2009) Sociality in Rancho La Brea *Smilodon*: arguments favour 'evidence' over 'coincidence'. Biol Lett 5:563–564

Kraglievich L (1928) Contribución al conocimiento de los grandes cánidos extinguidos de Sud América. An Soc Cient Argent 106:25–66

Kraglievich L (1930) Craneometría y clasificación de los cánidos sudamericanos, especialmente los argentinos actuales y fósiles. Physis 10:35–73

Kraglievich L (1934) La antigüedad Pliocena de las Faunas de Monte Hermoso y Chapadmalal, deducidas de su comparación con las que le precedieron y sucedieron. Imprenta El Siglo Ilustrado, Monte-video

Kraglievich JL (1952) El perfil geológico de Chapadmalal y Miramar. Provincia de Buenos Aires. Revista del Museo de Ciencias Naturales y Tradicional de Mar del Plata 1:8–37

Kraglievich JL, Olazábal AG (1959) Los prociónidos extinguidos del género *Chapalmalania* Ameghino. Rev Mus Arg Cienc Nat "B. Rivadavia" 6:1–59

Kraglievich JL, Reig OA (1954) Un nuevo prociónido del Plioceno de Las Playas (Provincia de Códoba). RAGA 9:209–231

Kraglievich L, Rusconi R (1931) Restos de vertebrados vivientes y extinguidos hallados por los señores ER Wagner y hermano en túmulos precolombinos de Santiago del Estero. Physis 10:229–241

Kurtén B (1968) Pleistocene mammals of Europe. Columbia University Press, New York

Kurtén B, Werdelin L (1990) Relationships between North and South American *Smilodon*. J Vert Paleontol 10:158–169

Langguth A (1975) Ecology and evolution in the South American canids. In: Fox MN (ed) The wild canids: their systematics, behavioral ecology and evolution. Van Nostrand Reinhold, New York, pp 192–206

Langguth A (1980) El origen del género *Speothos* y la evolución hacia *Speothos venaticus*. I Reunión Iberoamericana de Zoología de Vertebrados, Anales:587–600

Lessa G, Cartelle C, Faria H, Gonçalves P (1998) Novos achados de mamíferos carnívoros do Pleistoceno Final-Holoceno em grutas calcárias do estado da Bahia. Acta Geologica Leopoldensia 21:157–169

Lima ES, Jorge RSP, Dalponte JC (2009) Habitat use and diet of bush dogs, *Speothos venaticus*, in the northern Pantanal, Mato Grosso, Brazil. Mammalia 73:13–19

Linares OJ (1998) Mamíferos de Venezuela. Sociedad Conservacionista Audubon de Venezuela, Caracas

Lindsey EL, Seymour K (2015) "Tar Pits" of the western neotropics: paleoecology, taphonomy, and mammalian biogeography. Publciations of the Natural History Museum of Los Angeles County, Science Series 42:111–123

López Mendoza P, Mena Larraín F, Bostelmann E (2015) Presence of *Arctotherium* (Carnivora, Ursidae, Tremarctinae) in a precultural level of Baño Nuevo-1 cave (Central Patagonia, Chile). Estud Geol 71(2):1–8

Madurell Malapeira J, Alba DM, Moyà-Solà S, Aurell-Garrido J (2010) The Iberian record of the puma-like cat *Puma pardoides* (Owen, 1846) (Carnivora, Felidae). C R Acad Sci, Palevol 9:55–62

Marshall LG, Butler RF, Drake RE, Curtis GH, Tedford RH (1979) Calibration of the Great American interchange. Science 204:272–279

Martin FM (2013) Taphonomía y paleoecología de la transición Pleistoceno-Holoceno en Fuego-Patagonia. Interacción entre humanos y carnívoros y su importancia como agentes en la formación del registro fósil. Ediciones de la Universidad de Magallanes, Punta Arenas, pp 1–406

Massoia E (1992) Zooarqueología, I. Mammalia. In: Fernández J (ed) La Cueva de Haichol. Arqueología de los Pinares Cordilleranos del Neuquén, II. Anales de Arqueología y Etnología 1992:43–45, pp 447–505

McCall S, Naples V, Martin L (2003) Assessing behavior in extinct animals: was *Smilodon* social? Brain Behav Evol 61:159–164

McHenry CR, Wroe S, Clausen PD, Moreno K, Cunningham E (2007) Supermodeled sabercat, predatory behavior in *Smilodon fatalis* revealed by high-resolution 3D computer simulation. Proc Natl Acad Sci USA 104:16010–16015

Meachen-Samuels J, Van Valkenburgh B (2009) Forelimb indicators of prey-size preference in the Felidae. J Morphol 270:729–744

Méndez C, Barberena R, Reyes O, Nuevo Delaunay A (2014) Isotopic ecology and human diets in the forest-steppe ecotone, Aisén region, central-western Patagonia, Chile. Int J Osteoarchaeol 24:187–201

Mitchell KJ, Bray SC, Bover P, Soibelzon L, Schubert BW, Prevosti FJ, Prieto A, Martin F, Austin JJ, Cooper A (2016) Ancient mitochondrial DNA reveals convergent evolution of giant short-faced bears (Tremarctinae) in North and South America. Biol Let 12:20160062

Mones A, Rinderknecht A (2004) The first South American Homotheriini (Mammalia: Carnivora: Felidae). Comunicaciones Paleontológicas del Museo Nacional de Historia Natural y Antropología 35:201–212

Morales MM, Giannini NP (2014) Pleistocene extinctions and the perceived morphofunctional structure of the neotropical felid ensemble. J Mamm Evol 21:395–405

Morey D (2010) Dogs: domestication and the development of a social bond. Cambridge University Press, Cambridge

Morlo M, Peigné S, Nagel D (2004) A new species of Prosansanosmilus: implications for the systematic relationships of the family Barbourofelidae new rank (Carnivora, Mammalia). Zoolog J Linn Soc 1401:43–61

Moura Bubadué J, Cáceres N, dos Santos Carvalho R, Meloro C (2015) Ecogeographical variation in skull shape of South-American canids: abiotic or biotic processes? Evol Biol 43:145–159

Nordenskiöld E (1908) Ein neuer Fundort für Säugetierfossilien in Peru. Arkiv för Zoologi 4:1–22

Nyakatura K, Bininda-Emonds OR (2012) Updating the evolutionary history of Carnivora (Mammalia): a new species-level supertree complete with divergence time estimates. BMC Biol 10:1–12.

O'Leary MA, Bloch JI, Flynn JJ, Gaudin TJ, Giallombardo A, Giannini NP, Goldberg SL, Kraatz BP, Luo Z-X, Meng J, Ni X, Novacek MJ, Perini FA, Randall ZS, Rougier GW, Sargis EJ, Silcox MT, Simmons NB, Spaulding M, Velazco PM, Weksler M, Wible JR, Cirranello AL (2013) The placental mammal ancestor and the post K-Pg radiation of placentals. Science 339:662–667

Ochsenius C, Gruhn R (1979) Taima Taima: final report on the 1976 excavations. Monografías Científicas 3. Universidad Nacional Experimental Francisco de Miranda, Coro, Venezuela

Olivares I, Verzi D, Vucetich M (2012) Definición del género *Eumysops* Ameghino, 1888 (Rodentia, Echimyidae) y sistemática de las especies del Plioceno temprano de la Argentina central. Ameghiniana 49:198–216

Pascual R (1958) *Lyncodon bosei* nueva especie del Ensenadense. Un antecesor del huroncito patagónico. Revista del Museo de La Plata, Serie Paleontología 4:1–34

Paunero RS, Frank A, Skarbun F, Rosales G, Zapata G, Cueto M, Paunero MF, López R, Lunazzi N, Del Giogio M (2005). Arte rupestre en estancia la maría, meseta central de santa cruz: sectorización y contextos arqueológicos. Relaciones de la Sociedad de Antropología (n. s.) 30:1–26

Perea D, Manzuetti A, Ubilla M, Da Silva JS (2015) Nuevo félido (Mammalia, Carnivora) para la Fm. Sopas (Pleistoceno tardío) de Uruguay. XXIX Jornadas Argentinas de Paleontología de Vertebrados, Resúmenes, p 67

Poiré DG, de los Reyes M, Tineo D, Bona P, Perez LM, Vergani GD, González G, Reguero M (2015) La Angostura: Una nueva localidad fosilífera de vertebrados para la Formación Tariquía (Neogeno) en el Subandino de Bolivia. XXIX Jornadas Argentinas de Paleontología de Vertebrados, Resúmenes, pp 68–69

Politis G, Barrientos G, Scabuzzo C (2014) Los entierros humanos de Arroyo Seco 2. In: Politis G, Gutiérrez MA, Scabuzzo C (eds). Estado actual de las investigaciones en el sitio arqueológico Arroyo Seco 2, Partido de Tres Arroyos, Provincia de Buenos Aires, Argentina. Buenos Aires: Universidad Nacional del Centro de la Provincia de Buenos Aires, Facultad Ciencias Sociales, pp 329–392

Pomi LH, Prevosti FJ (2005) Sobre el status sistemático de *Felis longifrons* Burmeister, 1866 (Carnivora: Felidae). Ameghiniana 42:489–494

Prates L, Prevosti FJ, Berón M (2010) First records of prehispanic dogs in southern South America (Pampa-Patagonia, Argentina). Curr Anthropol 51:273–280

Prevosti FJ (2006a) New materials of Pleistocene cats (Carnivora, Felidae) from southern South America, with comments on biogeography and the fossil record. Geobios 39:679–694

Prevosti FJ (2006b) Grandes caánidos (Carnivora, Canidae) del cuaternario de la Republica Argentina: Sistemática, Filogenia, Bioestratigrafía y Paleoecología. PhD Thesis, Universidad Nacional de La Plata, La Plata, Argentina

Prevosti FJ (2010) Phylogeny of the large extinct South American canids (Mammalia, Carnivora, Canidae) using a "total evidence" approach. Cladistics 26:456–481

Prevosti FJ, Ferrero BS (2008) A Pleistocene giant river otter from Argentina: remarks on the fossil record and phylogenetic analysis. J Vert Paleontol 28:1171–1181

Prevosti FJ, Martin FM (2013) Paleoecology of the mammalian predator guild of the southern patagonia during the latest Pleistocene: ecomorphology, stable isotopes, and taphonomy. Quatern Int 305:74–84

Prevosti FJ, Palmqvist P (2001) Análisis ecomorfológico del cánido hipercarnívoro *Theriodictis platensis* (Mammalia, Carnivora) basado en un nuevo ejemplar del Pleistoceno de Sudamérica. Ameghiniana 38:375–384

Prevosti FJ, Pardiñas UFJ (2001) Variaciones corologicas de *Lyncodon patagonicus* (Carnivora, Mustelidae) durante el cuaternario. Mastozool Neotrop 8:21–39

Prevosti FJ, Pardiñas UFJ (2009) Comments on "The oldest South American Cricetidae (Rodentia) and Mustelidae (Carnivora): late Miocene faunal turnover in central Argentina and the Great American biotic interchange" by D.H. Verzi and C.I. Montalvo [Palaeogeography, Palaeoclimatology, Palaeoecology 267 (2008) 284–291]. Palaeogeogr Palaeoclimatol Palaeoecol 280:543–547

Prevosti FJ, Pardiñas UFJ (in press) The heralds: carnivores (Carnivora) and sigmodontine rodents (Cricetidae) in the Great American Biotic Interchange. In: Rosenberger AL, Tejedor MF (eds) Origins and evolution of Cenozoic South American mammals. Vertebrate Palebiology and Paleoanthropology Book Series, Springer, New York

Prevosti FJ, Pomi LH (2007) *Smilodontidion riggii* (Carnivora, Felidae, Machairodontinae): revisión sistemática del supuesto félido chapadmalalense. Revista del Museo Argentino de Ciencias Naturales (n.s.) 9:67–77

Prevosti FJ, Rincón AD (2007) A new fossil canid assemblage from the Late Pleistocene of northern South America: the canids of the Inciarte Asphalt Pit (Zulia, Venezuela), fossil record and biogeography. J Paleontol 81:1053–1065

Prevosti FJ, Schubert B (2013) First taxon date and stable isotopes (δ^{13}C, δ^{15}N) for the large hypercarnivorous South American canid *Protocyon troglodytes* (Canidae, Carnivora). Quatern Int 305:67–73

Prevosti FJ, Soibelzon LH (2012) The evolution of South American carnivore fauna: a paleontological perspective. In: Patterson B, Costa LP (eds) Bones, clones and biomes: the history and geography of recent neotropical mammals. University Chicago Press, Chicago, pp 102–122

Prevosti FJ, Vizcaíno S (2006) The carnivore guild of the Late Pleistocene of Argentina: Paleoecology and carnivore richness. Acta Palaeontol Pol 51:407–422

Prevosti FJ, Soibelzon LH, Prieto A, San Roman M, Morello F (2003) The southernmost Bear: *Pararctotherium* (Carnivoram Ursidaem Tremarctinae) in the latest Pleistocene of Southern Patagonia, Chile. J Vert Paleontol 23:709–712

Prevosti FJ, Bonomo M, Tonni EP (2004) La distribucion de *Chrysocyon brachyurus* (Illiger, 1811) (Mammalia: Carnivora: Canidae) durante el Holoceno en la Argentina: implicancias paleoambientales. Mastozool Neotrop 11:27–43

Prevosti FJ, Tonni EP, Bidegain JC (2009a) Stratigraphic range of the large canids (Carnivora, Canidae) in South America, and its relevance to Quaternary biostratigraphy. Quatern Int 210:76–81

Prevosti FJ, Ubilla M, Perea D (2009b) Large extinct canids from the Pleistocene of Uruguay: systematic, biogeographic and paleoecological remarks Large extinct canids from the Pleistocene of Uruguay: systematic, biogeogra. Hist Biol 21:79–89

Prevosti FJ, Turazzini GF, Chemisquy MA (2010) Morfología craneana en tigres dientes de sable: alometría, función y filogenia. Ameghiniana 47:239–256

Prevosti FJ, Santiago F, Prates L, Salemme M (2011) Constraining the time of extinction of the South American fox *Dusicyon avus* (Carnivora, Canidae) during the late Holocene. Quatern Int 245:209–217

Prevosti FJ, Forasiepi A, Zimicz N (2013) The evolution of the cenozoic terrestrial mammalian predator guild in South America: competition or replacement? J Mamm Evol 20:3–21

Prevosti FJ, Ramírez MA, Martin F, Udrizar Sauthier DE, Carrera M (2015) Extinctions in near time: new radiocarbon dates point to a very recent disappearance of the South American fox Dusicyon avus (Carnivora: Canidae) Biol J Linn Soc 116:704–720

Quintana CA (2001) Composición y cambios en la secuencia faunística. In: Mazzanti DL, Quintana CA (eds) Cueva Tixi: cazadores y recolectores de las Sierras de Tandilia Oriental. 1. Geología, Paleontología y Zoooarqueología. Publicación 1 del Laboratorio de Arqueología, Universidad Nacional de Mar del Plata, Mar del Plata, pp 37–64

Radinsky LB (1981) Evolution of skull shape in carnivores: 1. Representative modern carnivores. Biol J Linn Soc 15:369–388

Ramírez MA (2014) A new species of *Cerdocyon* (Carnivora, Canidae) from the Lujanian of Buenos Aires (Argentina). 4th International Paleontological Congress, Abstracts, p 790

Ramirez MA, Prevosti FJ (2014) Systematic revision of "*Canis*" *ensenadensis* Ameghino, 1888 (Carnivora, Canidae) and the description of a new specimen from the Pleistocene of Argentina. Ameghiniana 51:37–51

Ramírez MA, Prevosti FJ, Acosta A, Buc N, Loponte D (2015) On the presence of *Nasua* in the Buenos Aires province in the late Holocene. Rev Mus Arg Cienc Nat "B. Rivadavia", n.s. 17 (1):51–58

Reig OA (1952) Sobre la presencia de mustelidos mefitinos en la formación de Chapadmalal. Revista del Museo Municipal de Ciencias Naturales de Mar del Plata 1:45–51

Reig OA (1957) Un mustélido del género Galictis del Eocuartario de la provincia de Buenos Aires. Ameghiniana 1:33–47

Reig OA (1958) Notas para una actualización del conocimiento de la fauna de la Formación Chapadmalal. I. Lista faunistica preliminar. Acta Geológica Lilloana 2:241–253

Rincón AD (2006) A first record of the Pleistocene saber-toothed cat *Smilodon populator* Lund, 1842 (Carnivora: Felidae: Machairodontinae) from Venezuela. Ameghiniana 43:499–501

Rincón AD, Parra GE, Prevosti FJ, Alberdi T, Bell CJ (2009) A preliminary assessment of the mammalian fauna from the Pliocene-Pleistocene El Breal de Orocual Locality, Monagas state, Venezuela. Mus North Ariz Bull 65:593–620

Rincón AD, Prevosti FJ, Parras GE (2011) New saber-toothed cat records (Felidae: Machairodontinae) for the Pleistocene of Venezuela, and the Great American Biotic Interchange. J Vert Paleontol 31:468–478

Rodrigues S, Avilla L, Kugland De Azevedo SA (2016) Diversity and paleoenviromental significance of Brazilian fossil Galictis (Carnivora: Mustelidae). Hist Biol 28:907–912

Rodriguez SG, Soibelzon LH, Rodrigues S, Morgan CC, Bernardes C, Avilla L, Lynch E (2013) First record of *Procyon cancrivorus* (G. Cuvier, 1798) (Carnivora, Procyonidae) in stratigraphic context in the Late Pleistocene of Brazil. J S Amer Earth Sci 45:1–5

Rodrigues S, Avilla L, Soibelzon L, Bernades C (2014). Late pleistocene carnivores (Carnivora: Mammalia) from a cave sedimentary deposit in northern Brazil. Anais da Academia Brasileira de Ciências 86:1641–1655

Rusconi C (1932) Dos nuevas especies de mustélidos del piso ensenadense. "Grisonella hennigi" n. sp. et "Conepatus mercedensis pracecursor" subsp. n. Anales de la Sociedad Cientifica Argentina 113:42–45

Sakamoto M, Ruta M (2012) Convergence and divergence in the evolution of cat skulls: temporal and spatial patterns of morphological diversity. PLoS ONE 7(7):e39752

Sato JJ, Wolsan M, Prevosti FJ, D'Elía G, Begg C, Begg K, Hosoda T, Campbell KL, Suzuki H (2012) Evolutionary and biogeographic history of weasel-like carnivorans (Musteloidea). Mol Phylogenet Evol 63:745–757

Schiaffini MI, Martin GM, Giménez AL, Prevosti FJ (2013a) Distribution of *Lyncodon patagonicus* (Carnivora: Mustelidae): changes from the last glacial maximum to the present. J Mammal 94:339–350

Schiaffini MI, Gabrielli M, Prevosti FJ, Cardoso YP, Castillo D, Bo R, Casanave E, Lizarralde M (2013b) Taxonomic status of southern South American *Conepatus* (Carnivora: Mephitidae). Zool J Linn Soc 167:327–344

Schwartz M (1997) A history of dogs in the early Americas. Yale University Press, New Haven

Seymour KL (1983) The Felinae (Mammalia: Felidae) from the Late Pleistocene tar seeps at Talara, Peru, with a critical examination of the fossil and recent felines of North and South America. Unpublished MSc Thesis, University of Toronto, Toronto

Seymour KL (1993) Size change in North American Quaternary jaguars. In: Martin RA, Barnosky AD (eds) Morphological change in Quaternary mammals of North America. Cambridge University Press, Cambridge, pp 343–372

Seymour KL (1999) Taxonomy, morphology, paleontology and phylogeny of the South American small cats (Mammalia: Felidae). Unpublished Ph.D. Thesis, University of Toronto, Toronto

Shockey BJ, Salas-Gismondi R, Baby P, Guyot J-L, Baltazar MC, Huaman L, Clack A, Stucchi M, Pujos F, Emerson JM, Flynn JJ (2009) New Pleistocene cave faunas of the Andes of central Peru: radiocarbon ages and the survival of low latitude, Pleistocene DNA. Palaeontol Electron 12(3):1–15

Slater GJ, Van Valkenburgh B (2008) Long in the tooth: evolution of sabertooth cat cranial shape. Paleobiology 34:403–419

Soibelzon LH (2004) Revisión sistemática de los Tremarctinae (Carnivora, Ursidae) fósiles de América del Sur. Rev Mus Arg Cienc Nat "B. Rivadavia", n.s. 6:105–131

Soibelzon LH (2011) First description of milk teeth of fossil South American procyonid from the lower Chapadmalalan (late Miocene-early Pliocene) of "Farola Monte Hermoso", Argentina: paleoecological considerations. Palaontologische Z 85:83–89

Soibelzon LH, Prevosti FJ (2007) Los Carnívoros (Carnivora, Mammalia) terrestres del Cuaternario de América del Sur. In: Pons GX, Vicens D (eds) Geomorfologia Litoral i Quaternari. Homenatge a Joan Cuerda Barceló. Monografies de la Societat d'Història Natural de les Balears 14:49–68

Soibelzon LH, Prevosti FJ (2013) Fossils of South American land carnivores (Mammalia, Carnivora). In: Ruiz-García M, Shostell JM (eds) Molecular population genetics, evolutionary biology and biology conservation of neotropical carnivores. Nova Publishers, Nueva York, pp 509–527

Soibelzon LH, Schubert BW (2011) The largest known bear, *Arctotherium angustidens*, from the early Pleistocene pampean region of Argentina: with a discussion of size and diet trends in bears. J Paleontol 85:69–75

Soibelzon LH, Tarantini VB (2009) Estimación de la masa corporal de las especies de osos fósiles y actuales (Ursidae, Tremarctinae) de América del Sur. Rev Mus Arg Cienc Nat "B. Rivadavia", n.s. 11:243–254

Soibelzon E, Gasparini GM, Zurita AE, Soibelzon LH (2008) Análisis faunístico de vertebrados de las "toscas del Río de La Plata" (Buenos Aires, Argentina): un yacimiento paleontológico en desaparición. Rev Mus Arg Cienc Nat "B. Rivadavia", n.s. 10:291–308

Soibelzon LH, Gasparini GM, Soibelzon E (2010) Primer registro fósil de *Procyon cancrivorus* (G. Cuvier, 1798) (Carnivora, Procyonidae) en la Argentina. Rev Mex Cienc Geol 27:313–319

Soibelzon LH, Grinspan GA, Bocherens H, Acosta WG, Jones W, Blanco ER, Prevosti FJ (2014) South American giant short-faced bear (*Arctotherium angustidens*) diet: evidence from pathology, morphology, stable isotopes, and biomechanics. J Paleontol 88:1240–1250

Solé F, Smith T (2013) Dispersals of placental carnivorous mammals (Carnivoramorpha, Oxyaenodonta & Hyaenodontida) near the Paleocene-Eocene boundary: a climatic and almost worldwide story. Geologica Belgica 16(4):254–261

Stahl P (2003) The zooarchaeological record form Formative Ecuador. In: Raymond JS, Burger RL (eds) Archaeology of formative Ecuador. Dumbarton Oaks Research Library and Collection, Washington D.C., pp 175–212

Stahl PW (2013) Early dogs and endemic south American canids of the Spanish main. J Anthropol Res 69:515–533

Stahl P, Athens JS (2001) A high elevation zooarchaeological assemblage from the northern Andes of Ecuador. J Field Archaeol 28:161–176

Stucchi M, Salas-Gismondi R, Baby P, Guyot J-L (2009) A 6,000 + year-old specimen of a spectacled bear from an Andean cave in Peru. Ursus 20:63–68

Tarquini J, Morgan C, Soibelzon L, Toledo N (2015) Estimacion del tamaño corporal de los prociónidos (Mammalia, Carnivora) fósiles del "grupo *Cyonasua*" Reunión de Comunicaciones de la Asociación Paleontológica Argentina. Resúmenes 1:23–24

Tarquini J, Toledo N, Morgan CC, Soibelzon LH (2017) The forelimb of †*Cyonasua* sp. (Procyonidae, Carnivora): ecomorphological interpretation in the context of carnivorans. Earth Env Sci T R So 106:325–335

Tarquini J, Vilchez Barral MG, Soibelzon L (2016) Los prociónicos fósiles de América del Sur. Contribuciones del MACN 6:359–365

Tedford RH, Wang X, Taylor BE (2009) Phylogenetic systematics of the North American fossil Caninae (Carnivora: Canidae). Bull Amer Mus Nat Hist 325:1–218

Toledo MJ (2011) El legado lujanense de Ameghino: revisión estratigráfica de los depósitos pleistocenos-holocenos del valle del río Lujan en su sección tipo. Registro paleoclimático en la pampa de los estadíos OIS 4 al OIS 1. RAGA 68:121–167

Tonni EP, Cione AL, Figini A, Glaz D, Gasparini GM (2002) El "piso Aymará" de la región pampeana de la Argentina. Cronología radiocarbónica y paleontología. Ameghiniana 39 (3):313–319

Trejo V, Jackson D (1998) Cánidos patagónicos: identificacíon taxonómica de mandíbulas y molares del sitio arqueológico Cueva Baño Nuevo 1. Anales del Instituto de la Patagonia. Serie Ciencias Humanas 26:181–194

Trigo TC, Schneider A, De Oliveira TG, Lehugeur LM, Silveira L, Freitas TRO, Eizirik E (2013) Molecular data reveal complex hybridization and a cryptic species of neotropical wild cat. Curr Biol 23:2528–2533

Turner A, Antón M (1996) The giant hyaena, *Pachycrocuta brevirostris* (Mammalia, Carnivora, Hyaenidae). Geobios 29:455–468

Ubilla M, Perea D (1999) Quaternary vertebrates of Uruguay: a biostratigraphic, biogeographic and climatic overview. In: Rabassa J, Salemme M (eds) Quat S Am Antarct Peninsula 12:75–90

Van Valkenburgh B (1989) Carnivore dental adaptation and diet: a study of trophic diversity within guilds. In: Gittleman JL (ed) Carnivore behavior, ecology, and evolution, vol 1. Cornell University Press, New York, pp 410–436

Van Valkenburgh B (1990) Skeletal and dental predictors of body mass in carnivores. In: Damuth J, Macfadden BJ (eds) Body size in Mammalian paleobiology: estimation and biological implication. Cambridge University Press, Cambridge, pp 181–205

Van Valkenburgh B, Hertel F (1998) The decline of North American predators during the Late Pleistocene. In: Saunders JJ, Styles BW, Baryshnikov GF (eds) Quaternary paleozoology in the Northern Hemisphere. Illinois State Museum Scientific Papers 27, Springfield, pp 357–374

Van Valkenburgh B, Grady F, Kurtén B (1990) The Plio-Pleistocene cheetah-like cat Miracinonyx inexpectatus of North America. J Vert Paleontol 1:434–454

Verzi DH, Montalvo CI (2008) The oldest South American Cricetidae (Rodentia) and Mustelidae (Carnivora): late Miocene faunal turnover in central Argentina and the Great American Biotic Interchange. Palaeogeogr Palaeoclimatol Palaeoecol 267:284–291

Wang X (1993) Transformation from plantigrady todigitigrady: functional morphology of locomotion in *Hesperocyon* (Canidae: Carnivora). Amer Mus Novitates 3069:1–23

Wang X, Carranza-Castañeda O (2008) Earliest hog-nosed skunk, Conepatus (Mephitidae, Carnivora), from the early Pliocene of Guanajuato, Mexico and origin of South American skunks. Zool J Linn Soc 154:386–407

Wang X, Tedford RH (2008) Dogs: their fossils relatives & evolutionary history. Columbia University Press, New York

Wang X, Tedford RH, Taylor BE (1999) Phylogenetic systematics of the Borophaginae (Carnivora: Canidae). Bull Amer Mus Nat Hist 243:1–391

Werdelin L (1985) Small Pleistocene felines of North America. J Vertebr Paleontol 5:94–210

Werdelin L (1991) Pleistocene vertebrates from Tarija, Bolivia in the collections of the Swedish Museum of Natural History. In: Suarez-Soruco R (ed.) Fósiles y Fascies de Bolivia, 1-Vertebrados. Revista Técnica de YPFB 12(3–4):273–284

Werdelin L (1996) Carnivoran ecomorphology: a phylogenetic perspective. In: Gittleman JL (ed) Carnivore behavior, ecology, and evolution, vol 2. Cornell University Press, New York, pp 582–624

Werdelin L, Yamaguchi N, Johnson W, O'Brien S (2008) Phylogeny and evolution of cats (Felidae). In: Macdonald DW, Loveridge AJ (eds) Biology and conservation of wild felids. Oxford University Press, Oxford, pp 59–82

Wilson DE, Mittermeier RA (2009) Handbook of the mammals of the world, vol 1: Carnivores. Lynx Editions, Barcelona

Wolsan M (1993) Phylogeny and classification of early European Mustelidae (Mammalia: Carnivora). Acta Theriol 38:345–384

Woodburne M (2004) Late Cretaceous and Cenozoic Mammals of North America. Columbia University Press, New York

Wroe S, Argot C, Dickman C (2004) On the rarity of big fierce carnivores and primacy of isolation and area: tracking large mammalian carnivore diversity on two isolated continents. Proc Royal Soc Lond 271:1203–1211

Wroe S, Chamoli U, Parr WCH, Clausen P, Ridgely R, Witmer L (2013) Comparative biomechanical modeling of metatherian and placental saber-tooths: a different kind of bite for an extreme pouched predator. PLoS ONE 8(6):e66888

Zunino GE, Vaccaro OB, Canevari M, Gardner AL (1995) Taxonomy of the genus *Lycalopex* (Carnivora: Canidae) in Argentina. P Biol Soc Wash 108:729–747

Chapter 5
The Fossil Record of Mammalian Carnivores in South America: Bias and Limitations

Abstract South America has a rich fossil record that allows the reconstruction of the continental communities during the Cenozoic. Florentino Ameghino was one of the earliest advocates of a temporal sequence of faunas and biogeographic events, later refined by several authors (e.g., George G. Simpson, Rosendo Pascual, Bryan Patterson). This scheme is continually revised and improved by new faunal, systematic, and chronological studies. The fossil record is always incomplete, and many biases are recognized, some of them—the megabiases affect the interpretation of the global fossil record. For example, in South America, a megabias exists with respect to tropical areas, particularly before the Late Pleistocene. The SA fossil record contains large hiatuses between ages, with some ages being unconstrained by geochronological dates, while others are poorly sampled in terms of fossil recovery, faunal diversity, and identified localities. This form of bias which together with the differential duration of the South American Ages affects interpretation of the evolution of the continental fauna. In this chapter, we examine the spatial distribution of South American fossil localities, their frequency per age in the Cenozoic, and discuss the effect biases in the fossil record by means of a statistical approach.

Keywords Megabias · Taphonomy · Paleogeography · Paleoenvironments

5.1 Introduction

Florentino Ameghino (1854–1911) was one of the earliest researchers to make a substantial contribution to knowledge of SA mammalian faunas, their evolution, and biogeography. His work (e.g., Ameghino 1889, 1906) became the basis for investigations by other authors, for example George G. Simpson (1950, 1980), who recognized three "faunal strata," mainly characterized by biogeographic events (namely isolation, new lineages coming from Africa, and the Great American Biotic Interchange; see Chap. 2). A large volume of work begun in the mid-twentieth century by many different research teams (e.g., Simpson 1950, 1980; Pascual and Odreman Rivas 1971; Patterson and Pascual 1972; Marshall et al. 1977, 1981,

© Springer International Publishing AG 2018
F.J. Prevosti and A.M. Forasiepi, *Evolution of South American Mammalian Predators During the Cenozoic: Paleobiogeographic and Paleoenvironmental Contingencies*, Springer Geology, https://doi.org/10.1007/978-3-319-03701-1_5

1983, 1984, 1985; Stehli and Webb 1985; Pascual and Ortiz-Jaureguizar 1990; Cione and Tonni 1995, 1996, 1999, 2001, 2005; Marshall and Cifelli 1990; Webb 1991, 2006; Flynn and Swisher 1995; Kay et al. 1999; Flynn et al. 2003; Pascual 2006; Cerdeño et al. 2008; Tejedor et al. 2009; Tonni 2009; Madden et al. 2010; Woodburne 2010; Vizcaíno et al. 2012; Dunn et al. 2013; Goin et al. 2012, 2016; Deschamps et al. 2013; Tomassini et al. 2013; Wilf et al. 2013; Woodburne et al. 2014a, b; Cione et al. 2015) produced the synthesis that we have to date concerning the biostratigraphy, chronology, biogeography, and evolution of Cenozoic fossil associations in South America (see also Table 1.1).

This copious bibliography is an indication of the amount and richness of the South American fossil record as well as the volume of study. But the record is imperfect and has important systematic biases (e.g., Marshall and Cifelli 1990; Prevosti and Soibelzon 2012; Prevosti et al. 2013; Carrillo et al. 2015). The fossil record is not consistent through different parts of the Cenozoic and across different regions of the continent, creating a megabias (large-scale distortions caused by changes in the quality of the fossil record; Kowalewski and Flessa 1996; Behrensmeyer et al. 2000; Noto 2011; Benton 2015). In addition, large-scale discontinuities in the record and taphonomic conditions of fossil sites are linked to differential intensity of sampling and imprecise limits for several ages (Marshall and Cifelli 1990; Prevosti and Soibelzon 2012; Prevosti et al. 2013).

In this chapter, we provide a short review of the quality of the Cenozoic continental fossil record of South America, with particular reference to the context of change in diversity of sparassodonts and carnivorans. In this and the following chapter, we test hypotheses about competition between clades and extinction and quantitatively evaluate if the bias in the fossil record has an impact on hypotheses.

5.2   Limitations and Bias of the Cenozoic Continental Fossil Record of South America

The biases examined are general problems that affect the fossil record (e.g., Marshall and Cifelli 1990; Kowalewski and Flessa 1996; Behrensmeyer et al. 2000; Noto 2011; Benton 2015) and in particular estimations of diversity, first and last appearances, and the precise calibration of different chronological parameters (e.g., Marshall and Cifelli 1990; Maas et al. 1995; Foote 2000; Prevosti and Soibelzon 2012; Prevosti et al. 2013; Benton 2015).

5.2.1   Hiatuses and Definition of Ages

Marshall and Cifelli (1990) discussed the importance of hiatuses, particularly in the Paleogene record. Since that time, new information has accumulated (e.g., Flynn et al. 2003; Cerdeño et al. 2008; Tejedor et al. 2009; Madden et al. 2010; Dunn et al. 2013; Clyde et al. 2014; Woodburne et al. 2014a, b), but hiatuses and unconstrained age limits remain in the South American fossil record (Tables 1.1, 5.1).

Major hiatuses in the Paleogene exist between most ages. In the case of the Peligran and the Itaboraian, the Riochican and the Vacan, and within the Casamayoran (Table 1.1) for example, hiatuses are longer than the time span of their corresponding ages (Dunn et al. 2013; Clyde et al. 2014; Woodburne et al. 2014a, b). New mammal associations partially fill those gaps and potentially represent new biochronological units (Tejedor et al. 2009; Madden et al. 2010).

The hiatus problem is less severe in the Neogene, but gaps of about 2 Ma exist, for example, between the Colhuehuapian and Santacrucian (or "Pinturan," Vizcaíno et al. 2012; Perkins et al. 2012).

For the Neogene, the most relevant issues are the lack of definition for some ages (e.g., Colloncuran, Friasian, Mayoan, Huayquerian) (e.g., Prevosti et al. 2013; Cione et al. 2015) and the absence of radiometric constraints.

For the late Miocene, there is an issue with the definition of the Huayquerian, one of the most extensive age units, repeatedly used in South America, from Venezuela to Argentina, and one of the most relevant in terms of the biochronology of the GABI, since this age records the earliest Holarctic immigrants in South America. The Huayquerian age was defined in the badlands of the Huayquerías of Mendoza (Mendoza, Argentina; De Carles 1911; Rovereto 1914), first used by Kraglievich (1934) imploying the concept of faunal association and then by Simpson (1940) as part of his scheme of South American Stages. However, the Huayquerías Formation in its type locality has provided less than a dozen published species (Forasiepi et al. 2014) and is poorly constrained, with a radiometric date of about 5.8 Ma in the upper part of the sequence (Yrigoyen 1994) from a re-deposited ash. Consequently, there are different conceptions in the literature about the Huayquerian that are not linked to the type locality and fauna found at the Huayquerías of Mendoza. Furthermore, the age was divided into a lower and upper Huayquerian intervals. The first was defined on the basis of the *Macrochorobates scalabrinii* biozone from an innominate lithostratigraphic unit exposed in the lower valley of Chasicó Creek, Buenos Aires (Tonni et al. 1998). The second was based on four biochronological units (Verzi et al. 2008) from the Cerro Azul, Irene, and Epecuén formations exposed in eastern La Pampa and southwestern Buenos Aires provinces (Cione et al. 2015).

In an attempt to constrain the age of the Huayquerian, Cione et al. (2015) correlated its fauna with the "Mesopotamiense" in Entre Rios, Argentina (Brandoni and Noriega 2013), constrained by a date of 9.47 Ma obtained for the lower Paraná Formation (Pérez 2013). This sets the age of the Huayquerian between ca. 8.2 and 5.9 Ma. Based on the rodent record of the Cerro Azul Formation and deposits in the southwest of Buenos Aires Province (e.g., Verzi et al. 2008; Verzi and Montalvo 2008), other authors placed the Huayquerian between ca. 6 and ca. 5.3 Ma, although without radiometric or paleomagnetic data (see Prevosti and Pardiñas 2009 for a critique of this scheme). More recently, a date of 5.28 Ma was determined for

Table 5.1 Diversity of predators (carnivorans and sparassodonts) number of fossil localities, time span, and Lazarus taxon for the South American Ages

Age	Number of sites	Time span (Ma)	Geographic area (km²)	Predator diversity	Predator Lazarus	Spa. diversity	Spa. Lazarus	Car. diversity	Car. Lazarus
Tiupampan	1	0.7	175	1	0	1	0	0	0
Peligran	1	1.7	175	0	1	0	1	0	0
Itaboraian	2	1.5	21,116	1	0	1	0	0	0
Riochican	1	2	289	0	1	0	1	0	0
Casamayoran	9	10	19,774	9	0	9	0	0	0
Mustersan	3	2.5	234,294	3	0	3	0	0	0
Tinguirirican	1	2.3	5314	0	0	0	0	0	0
Deseadan	11	5.2	5,290,846	8	0	8	0	0	0
Colhuehuapian	5	0.9	270,261	7	0	7	0	0	0
Santacrucian	26	2	348,695	11	0	11	0	0	0
Friasian	2	0.8	370	4	3	4	3	0	0
Colloncuran	6	1.5	200,582	3	1	3	1	0	0
Laventan	14	2	42,961	5	0	5	0	0	0
Mayoan	1	1.8	350	0	1	0	1	0	0
Chasicoan	2	1	291,880	4	0	4	0	0	0
Huayquerian	12	3.72	3,578,542	11	0	7	0	4	0
Montehermosan	1	1.28	29,875	3	1	2	1	1	0
Chapadmalalan	9	1	3,174,916	5	0	2	0	3	0
Barrancalobian	1	0.1	0.65	0	2	0	0	0	2
Vorohuean	3	0.2	11,766	3	1	0	0	3	1
Sanandresian	3	0.92	110,253	2	2	0	0	2	2
Ensenadan	37	1.38	1,412,918	22	0	0	0	22	0

(continued)

Table 5.1 (continued)

Age	Number of sites	Time span (Ma)	Geographic area (km²)	Predator diversity	Predator Lazarus	Spa. diversity	Spa. Lazarus	Car. diversity	Car. Lazarus
Bonaerian	8	0.274	88,020	1	6	0	0	1	6
Lujanian	108	0.119	15,871,045	36	0	0	0	36	0
Platan	135	0.005508	15,528,203	23	7	0	0	23	7

Spa. = sparassodonts; Car. = carnivorans; Ma. = millions of years

the upper Huayquerian levels studied by Verzi and Montalvo (Schultz et al. 2006; Tomassini et al. 2013; Cione et al. 2015) close to the Atlantic coast of Buenos Aires Province, which is a slightly younger upper age limit.

The immediately preceding Chasicoan Age has only one date of 9.24 Ma (Schultz et al. 2006) but its limits are otherwise not constrained.

The Pliocene Montehermosan to Marplatan Ages were defined in the coastal region of Buenos Aires Province. The Montehermosan is based on the *Eumysops laeviplicatus* biozone (Tomassini et al. 2013 = *Trigodon gaudryi* and *Neocavia depressidens* biozones of Cione and Tonni 2005) and at its type locality is neither dated nor constrained. Tomassini et al. (2013) suggested that the Monte Hermoso Formation was deposited over a few hundred thousand years, and point to the existence of hiatuses between the Montehermosan, Huayquerian, and Chapadmalalan Ages. The Chapadmalalan is based on the *Paraglyptodon chapadmalensis* biozone and has a radiometric date of age of 3.3 Ma at the top of its type unit, the Chapadmalal "Formation" (Schultz et al. 1998), three meters below the contact with the overlying Barranca de Los Lobos "Formation," indicating that the Chapadmalalan upper limit is at least younger than 3.3 Ma. The Marplatan includes three Subages, namely the Barrancalobian, Vorohuean, and Sanandresian, based on the *Platygonus scagliai*, *Akodon lorenzinii*, and *Ctenomys chapalmalensis* biozones, respectively (Cione and Tonni 1995, 1999; Cione et al. 2015). There are no confidents limits defined for any of these ages. But a recent paleomagnetic study (Rico and Bidegain 2013) suggested that the upper limit of the Vorohuean Subage is ca. 2.6 Ma, and that Sanandresian is between 1.8 and 2.6 Ma. Recently, Isla et al. (2015) questioned the mammal association used for the definition of the Barrancalobian Subage, suggesting that it could represent a mixture of Chapadmalalan and younger faunas. This challenging argument should be further evaluated.

The ages of the Quaternary South American Ages and biozones were also defined in the Pampean Region. Time constraints, particularly for the Ensenadan, Bonaerian and Lujanian, are tentative (Cione and Tonni 1995, 1999; Prevosti et al. 2009; Cione et al. 2015).

In short, although there is not a continuous sedimentary record in most sites of the Pampean Region, the late Miocene-Pleistocene SA Stages/Ages—with the exception of the Huayquerian—were defined in this area. Most sites are geographically isolated with few datable elements, which lead to uncertainty, in particular to the time span represented by the units. More continuous sequences with classical fossil sites are present in western Argentina (e.g., Catamarca Province; Riggs and Patterson 1939; Marshall and Patterson 1981; Esteban et al. 2014). Independent biozones have been recently defined and constrained by several radiometric dates and magnetostratigraphic data (Reguero and Candela 2011; Esteban et al. 2014), but these schemes are difficult to correlate with the biostratigraphic scheme of the Pampean Region. A recent proposition suggests avoiding the use of "ages" in favor of using the international timescale (Brandoni 2013; Esteban et al. 2014), but this is only possible when accurate independent methods of dating are available and the faunal ages are precisely delimited.

5.2.2   Age Time Span

Durations of SA Ages are extremely unequal. Some ages comprise several million years (ca. 5 Ma for the Deseadan), while others are less than 200,000 years (ca. 120 ka for the Lujanian and less than 10 ka for the Platan) (Tables 1.1, 5.1, and Fig. 5.1). Increasing the duration of an age naturally increases the number of fossil sites and finds with a potential consequent increase in diversity (Marshall and Cifelli 1990; Maas et al. 1995; Palombo et al. 2009; Figueirido et al. 2012; Prevosti and Soibelzon 2012). However, this effect has been recently statistically tested for SA associations with studies focused on the carnivoran fossil record since the late Miocene to Present (number of species against time span of each age; Prevosti and Soibelzon 2012), and on the sparassodont and carnivoran fossil record for the entire Cenozoic (Prevosti et al. 2013). Despite the inequality of the duration of the ages, these studies produced non-significant correlations, suggesting that the major pattern of diversity through time has not been strongly affected by fossil recovery biases.

The updated dataset of Prevosti et al. (2013) (Table 5.1) provides a similar result (Spearman rank correlation) whether the analysis was done with the total sample (R: −0.047, p 0.825), or for only sparassodonts (R: 0.268, p 0.283), or only carnivorans (R: −0.134, p 0.713) (Fig. 5.1). This relationship was confirmed using Quantile Regression and the Durbin–Watson test that indicated the absence of autocorrelation in the data.

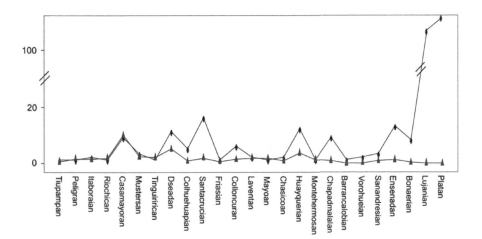

Fig. 5.1   Number of localities (a proxy of sampling effort) and time span of each South American Age. Black diamonds and line: number of localities; blue triangles and line: time span (millions of years)

5.2.3 Geographic Bias

The SA fossil record has a strong geographic megabias against tropical or low-latitude areas (e.g., Pascual and Odreman Rivas 1971; Patterson and Pascual 1972; Marshall et al. 1982; Marshall and Cifelli 1990; Prevosti and Soibelzon 2012; Carrillo et al. 2015; Goin et al. 2016). This bias is more extreme for the Paleogene and early–middle Miocene, because other than a few exceptions (e.g., Campbell 2004; Antoine et al. 2015), fossil sites are limited to high-latitude areas, especially in Patagonia (e.g., Pascual and Odreman Rivas 1971; Patterson and Pascual 1972; Marshall et al. 1982; Prevosti and Soibelzon 2012; Carrillo et al. 2015; Goin et al. 2016; Figs. 3.1, 4.1 and 5.2).

In contrast, post-middle Miocene fossil sites are generally found north of Patagonia. The distribution of late Miocene Huayquerian outcrops is wider, but still there are few records from lowland tropical South America or Patagonia. In this context, Prevosti and Soibelzon (2012) demonstrated that for the late Huayquerian–Platan time span, localities with fossil carnivorans are mostly restricted to the southern half of the continent and only since the Late Pleistocene (Lujanian) is there a wider coverage of the continent, including tropical and low-latitude areas. Similarly, Carrillo et al. (2015), in an analysis that included the complete faunal lists from 13 middle Miocene to Pleistocene associations, detected a differentiation between temperate and tropical faunas with a megabias toward higher latitudes and younger ages.

In order to test this megabias, we analyze the correlation between the geographic coverage of each biochronological unit vs its taxonomic diversity by considering the number of carnivorous species, both sparassodont and SA carnivoran. We

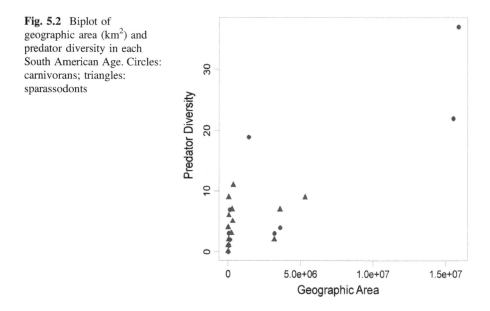

Fig. 5.2 Biplot of geographic area (km^2) and predator diversity in each South American Age. Circles: carnivorans; triangles: sparassodonts

estimated the geographic area covered by each SA Age, using localities that are the furthest apart localities in the west, east, south, and north, and applying the software gvSIG (http://www.gvsig.org); Table 5.1; raw data available on request). For ages with only one or two fossil localities, we duplicated or triplicated their geographic coordinates, and their first decimal was randomly changed. The analysis of these data indicates that there is a positive and significant correlation between the area covered by the sites of each age and the species richness of the whole sample (sparassodonts + carnivorans; Spearman R: 0.829, $p < 0.0000004$), only sparassodonts (Spearman R: 0.688, $p < 0.0016$), and only carnivorans (R: 0.844, $p < 0.0022$). These results indicate that ages covering larger geographic areas produced larger diversities (Fig. 5.2). This relationship was also confirmed using Quantile Regression (whole sample, sparassodonts + carnivorans: r^2 0.686, $p = 0.00342$; carnivoran sample: r^2: 0.687, $p = 0.015$), but not for the sparassodont sample, which gave a non-significant result (r^2: 0.126, $p = 0.09530$). The Durbin–Watson tests were not significant, indicating the absence of autocorrelation in the data.

5.2.4   Sampling

Differences in sampling between ages is an issue for a wide range of analyses, since it is likely that well-sampled ages provide larger diversities than those poorly sampled (e.g., Benton et al. 2000). Several proxies were used to evaluate the impact of sampling on the diversity counts: number of fossil localities, number of geological formations, geological exposure area, and skeletal completeness (e.g., Figueirido et al. 2012; Benton 2015; Cleary et al. 2015), but these measurements are also affected by differential sampling efforts, taphonomic processes, fossil site exposures, and specimen availability and/or taxonomic work. A non-significant relationship between diversity and one of these sampling proxies cannot completely rule out the presence of a bias, but a significant correlation may be interpreted as evidence of bias.

The sampling bias of the SA fossil record was tested by Prevosti and Soibelzon (2012) and Prevosti et al. (2013) using the nonparametric Spearman rank correlation between the diversity of carnivorans and sparassodonts and the number of localities per age (see also Marshall and Cifelli 1990). The results demonstrated the presence of a strong and significant correlation, congruent with the presence of a sampling bias, when carnivorans, sparassodonts, and both groups together were considered. The "range-through taxa" (="Lazarus taxa") were included in the tests (Fig. 5.3).

A new analysis of the data (Table 5.1; Figs. 5.1, 5.4) supports previous results, with a positive correlation between the number of sites and diversity for the SA Ages, whether the correlation is tested with the whole sample: sparassodonts + carnivorans (Spearman R: 0.817, $p < 0.0000006$), only with sparassodonts (Spearman R: 0.745, $p < 0.0004$), or only with carnivorans (Spearman R: 0.893, $p < 0.0005$). These results indicate that better-sampled ages provide larger

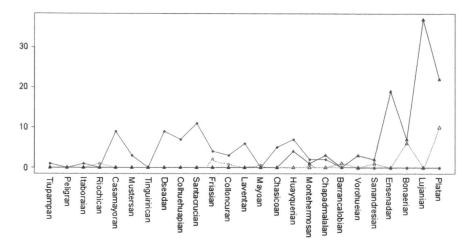

Fig. 5.3 Diversity of predators (number of taxa) and number of Lazarus taxa in each South American Age. Red circles and lines: sparassodonts; blue triangles and lines: carnivorans; solid lines: diversity; broken lines: Lazarus taxa

diversities. A relationship was also confirmed using Quantile Regression and the Durbin–Watson test, indicating lack of autocorrelation in the data. Consequently, the low diversity found for most of the Paleogene, and Colloncuran, Friasian, Chasicoan, Montehermosan, Marplatan, and Bonaerian Ages could be explained by this sampling bias (Figs. 5.1 and 5.4). "Lazarus taxa" showed a low negative relationship against diversity (Spearman R between -0.177 and -0.314), but this is non-significant ($p > 0.05$).

The number of localities and geographic areas covered by each age is strongly and significantly correlated, especially when considering the whole sample (sparassodonts + carnivorans) or the carnivoran sample alone (Spearman R: 0.84 and $p < 0.0000002$, and Spearman R: 0.84 $p < 0.0022$, respectively). The sparassodont sample had a lower but still significant correlation between these variables (Spearman R: 0.77, $p < 0.000185$). This high positive correlation indicates that geographic area and number of localities are redundant. In this sense, better-sampled ages will have more localities and a wider geographic covered area than the less sampled.

Because there is a clear recovery bias in the fossil record, we examined how this has affected observed sparassodont and carnivoran diversity during the late Miocene and Pliocene, which is the period of crucial relevance to understand the demise of sparassodonts. Using re-sampling techniques (Manly 1997), the sampling effort was tested with the number of localities and specimens as proxies, of each clade by age (Tables 5.1 and 5.2). We reduced the sample of the better-known associations (Santacrucian, the combination of Friasian to Chasicoan, and Huayquerian for sparassodonts; Ensenadan for carnivorans) to test if these approximate the sampling effort observed for the Huayquerian to Chapadmalalan

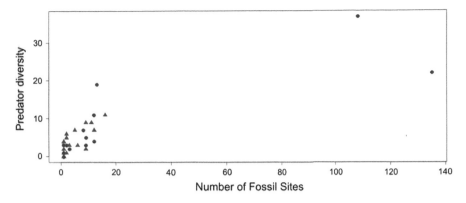

Fig. 5.4 Biplot of the number of localities and the diversity of predators (number of species) in each South American Age. Circles: carnivorans; triangles: sparassodonts

for sparassodonts and Huayquerian to Sanandresian for carnivorans. For sparassodonts, the Huayquerian was also used as a proxy to test the diversity of the Montehermosan and Chapadmalalan. If the diversity displayed by the fossil record is significantly different from the re-sampling exercise ($p < 0.05$), we can assume that difference in the sampling effort does not explain the low diversity observed for the late Miocene–Pliocene and should indicate a real pattern.

The result of our analysis demonstrated that the observed diversity of sparassodonts for the Huayquerian is lower than expected (except when it is compared with the Santacrucian) and for the Chapadmalalan (except when it is compared with the Huayquerian using the specimens as proxy for the sampling effort). In contrast, the difference in diversity in the Montehermosan is lower and non-significant (except when it is compared with the Santacrucian using the localities as proxy for the sampling effort). The observed carnivoran diversity is significantly lower during

Table 5.2 Number of specimens of carnivorans and sparassodonts during the Santacrucian–Ensenadan

Age	Carnivorans	Sparassodonts
Santacrucian	0	220
Friasian	0	8
Colloncuran	0	5
Laventan	0	26
Mayoan	0	0
Chasicoan	0	7
Huayquerian	16	31
Montehermosan	4	2
Chapadmalalan	18	9
Barrancalobian	0	0
Vorohuean	3	0
Sanandresian	2	0
Ensenadan	189	0

the Huayquerian, Montehermosan, and Chapadmalalan. But during the Vorohuean and Sanandresian, the difference in diversity is only significant when localities are used as proxy. Using the number of localities and re-sampling the Santacrucian and Huayquerian faunas, we calculated a median number of sparassodonts in 2–4, 3–6, 4–6, and 4–10 species bins for the Barrancalobian, Vorohuean, Sanandresian, and Ensenadan, respectively. Conversely, the median random expected number of carnivorans for the Barrancalobian is four species, twice the number of ghost lineages recorded in this Subage.

Finally, we examined the situation considering three carnivorans (*Conepatus altiramus, Smilodontidion riggii,* and *"Felis" pumoides*) with dubious Chapadmalalan stratigraphic procedence (see Chap. 4), as a simulation of potential new records that could increase the diversity of carnivorans for that age. However, we understand that is not possible to corroborate that these fossil were effectively recovered from Pliocene beds (Berman 1994; Cione and Tonni 1995, 2001, 2005, Prevosti and Soibelzon 2012). In only considering a diversity of six carnivores, the result is marginally non-significant, suggesting that the observed low diversity is a product of taphonomic biases. However, removing any one of these taxa results in a significant difference from random expectation.

5.3   Discussion and Conclusions

The review of several potential biases in the South American fossil record high-lights the limitations and complications in interpretation of diversity and related variables. Important issues regarding the SA biochronological scheme and major biases are the lack of definition for some units and lack of radiometric constraints for others, the presence of large hiatuses in the biochronological sequence, especially for the Paleogene, the difference in interval length, from the Deseadan covering ca. 5 Ma to the Platan covering less than 10 ka, a geographic bias in the fossil record disfavoring the tropics, and differential sampling.

In this context, sampling biases are probably a major reason for the difference between the high diversity and ecological disparity of the Santacrucian Age in comparison to the low values recovered for the Chasicoan. A relevant issue is to determine whether the long interval between the last record of the large hyper-carnivorous Sparassodonta in the Chapadmalalan and the first records of the large hypercarnivorous Carnivora (Felidae and some clades of Canidae) in the Ensenadan (Fig. 4.2) is a real pattern or a product of the bias of the record (Prevosti et al. 2013).

As was mentioned in Chapter 4, small hypercarnivorous carnivorans (e.g., *Galictis* spp.) are known since the Vorohuean but are more diverse since the Ensenadan. Foxes are known since the Vorohuean but have a better fossil record during the Ensenadan–Lujanian Ages. On the other hand, sabertooth cats and other large hypercarnivorous placentals (e.g., canids, *Puma, Panthera*) appear in the SA fossil record during the Ensenadan and are the most frequently recovered carnivorans during the last 1.8 Ma. The absence of large hypercarnivorous carnivorans in the Chapadmalalan–Marplatan could be due a strong bias against medium- and large-sized carnivorans, a conclusion congruent with the recent suggestion that the glyptodont *Panochthus* is a Lazarus taxon during the Marplatan (Zamorano et al. 2014; Cione et al. 2015; see also Isla et al. 2015). However, medium- and large-sized mammals, for example, equids, other glyptodonts, ground sloths, mesotheriids, and one large carnivoran (*Chapalmalania*), are recorded for the Marplatan (Cione and Tonni 2005; Cione et al. 2015). In this context, we expect that if *Smilodon* was present before the Ensenadan in the Pampean Region, it should have been found together with mustelids and foxes. This reasoning could also be applied to justify the absence of *Thylacosmilus* in the Marplatan. Additionally, carnivorous didelphimorphians are also recorded in the Barrancalobian and Vorohuean (Zimicz 2014) and, since they have small body sizes, this suggests that small hypercarnivore sparassodonts should be found in the Marplatan if they were not extinct. New collections with controlled provenance are needed to test the presence of this type of bias.

The Chapadmalalan Age is well known, particularly through the very rich outcrops of the Chapadmalal "Formation," which have been extensively explored for more than a century (see Cione and Tonni 1995). Several specimens of *Cyonasua*, some of *Chapalmalania* and *Thylacosmilus*, one specimen of *Borhyaenidium*, and "terror birds" have been collected here (see Chaps. 3 and 4; Tambussi and Degrange 2011; Degrange et al. 2015). The possible presence of felids in the Chapadmalalan has been refuted (see Prevosti 2006; Prevosti et al. 2006; Prevosti and Pomi 2007), and the presence of mephitids (*Conepatus*) is dubious (Cione and Tonni 1995; Woodburne et al. 2006; Woodburne 2010). The large sampling effort and the amount of taxonomic and biostratigraphic work on the Chapadmalal "Formation" indicate that the mammalian diversity would not increase significantly with new fossil collections. Sparassodont and carnivoran diversity is presently very low in the Chapadmalalan, but this does not appear to be related to a bias in the fossil record (Table 5.1). Additionally, several key relevant Ages (i.e., Chapadmalalan, Ensenadan–Recent) are well represented and well sampled, locally or regionally. Consequently, we interpret that the low diversity and narrow morphological disparity of carnivorans during the Huayquerian–Sanandresian is not an artifact of the sampling effort (see also Prevosti and Soibelzon (2012), and similarly the high diversity and broad disparity observed since the Ensenadan is a real pattern.

Our re-sampling analysis (using number of fossil localities and specimens as proxies) tested the impact of the sampling effort on the observed diversity. The re-sampling analysis indicates that the observed carnivoran diversity is significantly lower than random expectation for the Huayquerian–Chapadmalalan, something that suggests that the observed diversity is a real pattern and not a product of a low sample size for these Ages. The same is the case for sparassodonts for the Chapadmalalan, and less certain for the Huayquerian, but the results for the Montehermosan are dubious. As happens with the Montehermosan, the results about carnivoran diversity in the Vorohuean and Sanandresian are also ambiguous because there are fewer than three localities or predator specimens (2–4) and consequently the diversity could increase with larger samples. In addition, it is not possible to exclude the presence of other biases obscuring the real diversity (such as systematic biases against specific body sizes, ecological groups, or taxa). However, in the absence of other evidence, we conclude that the biases in the fossil record are not enough to explain the low diversity observed during the late Miocene–Pliocene.

In summary, big structural issues and biases are relevant to interpreting the SA biochronological framework and particularly the Paleogene ages. With the available data, it is not possible to definitively exclude the possibility that taphonomic and sampling biases mask a significant overlap in time between ecologically equivalent sparassodonts and carnivorans. However, in light of the arguments and statistical predictions discussed, it is probable that the lack of temporal and ecological overlap actually reflects the real pattern.

References

Ameghino F (1889) Contribución al conocimiento de los mamíferos fósiles de la República Argentina. Actas Acad Nac Cs Córdoba, vol 4. Buenos Aires

Ameghino F (1906) Les formations sedimentaires du Crétacé Supèrieur et du Tertiaire de Patagonie, avec un paralieleentre leurs faunes mammalogiques et celles de l'ancien continente. Anal Mus Nac Hist Nat Buenos Aires 3:1–568

Antoine PO, Abello MA, Adnet S, et al. (2015). A 60-million year Cenozoic history of western Amazonian ecosystems in Contamana, eastern Peru. Gondwana Res. doi: 10.1016/j.gr.2015.11.001

Behrensmeyer AK, Kidwell SM, Gastaldo RA (2000) Taphonomy and paleobiology. Paleobiology 26:103–144

Benton MJ (2015) Palaeodiversity and formation counts: redundancy or bias? Palaeontology 58:1003–1029

Benton MJ, Wills MA, Hitchin R (2000) Quality of the fossil record through time. Nature 403:534–537

Berman WD (1994) Los carnívoros continentales (Mammalia, Carnivora) del Cenozoico en la provincia de Buenos Aires. Unpublished Ph.D. Thesis, Universidad Nacional de La Plata, La Plata

Brandoni D (2013) Los mamíferos continentales del "Mesopotamiense" (Mioceno tardío) de Entre Ríos, Argentina. Diversidad, edad y paleogeografía. In: Brandoni D, Noriega JI (eds) El Neógeno de la Mesopotamia Argentina, Publicación especial APA, vol 14, pp179–191

Brandoni D, Noriega JI (2013) El Neógeno de la Mesopotamia Argentina. Publicación Especial APA 14. Asociación Paleontológica Argentina, Buenos Aires

Campbell KE Jr (ed) (2004) The Paleogene mammalian fauna of Santa Rosa, Amazonian Perú. Nat Hist Mus Los Angeles County, Sci Ser 40, Los Angeles

Carrillo JD, Forasiepi AM, Jaramillo C et al (2015) Neotropical mammal diversity and the Great American Biotic Interchange: spatial and temporal variation in South America's fossil record. Front Genet 5:1–11

Cerdeño E, López G, Reguero M (2008) Biostratigraphical considerations on the Divisaderan faunal assemblage. J. Vert. Paleontol 28:574–577

Cione AL, Tonni EP (1995) Chronostratigraphy and " Land-Mammal Ages" in the Cenozoic of Southern South America: principles, practices, and the " Uquian" problem. J Paleontol 69:135–159

Cione AL, Tonni EP (1996) Reassesment of the Pliocene-Pleistocene continental time scale of southern South America. Correlation of the Chapadmalalan with Bolivian sections. J South Am Earth Sci 9:221–236

Cione AL, Tonni EP (1999) Biostratigraphy and chronological scale of uppermost Cenozoic in the Pampean area, Argentina. In: Tonni EP, Cione AL (eds) Quaternary vertebrate palaeontology in South America, vol 12. Quat S Am Antarct Peninsula, pp. 23–52

Cione AL, Tonni EP (2001) Correlation of Pliocene to Holocene southern South American and European vertébrate-bearing units. B Soc Paleontol Ital 40:167–173

Cione AL, Tonni EP (2005) Bioestratigrafía basada en mamíferos del Cenozoico superior de la provincia de Buenos Aires, Argentina. In: de Barrio RE, Etcheverry RO, Caballé MF, Llambías E (eds) Geología y Recursos Minerales de la Provincia de Buenos Aires, vol 11. Relatorio del XVI Congreso Geológico Argentino, pp 183–200

Cione AJ, Gasparini GM, Soibelzon E, Soibelzon LH, Tonni EP (2015) The Great American Biotic Interchange: a South American perspective. Springer Earth System Sciences, Dordrecht

Cleary TJ, Moon BC, Dunhill AM, Benton MJ (2015) The fossil record of ichthyosaurs, completeness metrics and sampling biases. Palaeontology 58:521–536

Clyde WC, Wilf P, Iglesias A, Slingerland RL, Barnum T, Bijl PK, Bralower TJ, Brinkhuis H, Comer EE, Huber BT, Ibañez-Mejia M, Jicha BR, Krause JM, Schueth JD, Singer BS, Raigemborn MS, Schmitz MD, Sluijs A, Mdel Zamaloa C (2014) New age constraints for the Salamanca Formation and lower Río Chico Group in the western San Jorge Basin, Patagonia, Argentina: implications for K/Pg extinction recovery and land mammal age correlations. Geol Soc Am Bull 3–4:289–306

De Carles E (1911) Ensayo geológico descriptivo de las Guayquerías del Sur de Mendoza (Dep. de San Carlos). An Mus Nac Buenos Aires 22:77–95

Degrange FJ, Tambussi CP, Taglioretti ML et al (2015) A new Mesembriornithinae (Aves, Phorusrhacidae) provides new insights into the phylogeny and sensory capabilities of terror birds. J Vertebr Paleontol 35:e912656

Deschamps CM, Vucetich MG, Montalvo CI, Zárate MA (2013) Capybaras (Rodentia, Hydrochoeridae, Hydrochoerinae) and their bearing in the calibration of the late Miocene-Pliocene sequences of South America. J South Am Earth Sci 48:145–158

Dunn RE, Madden RH, Kohn MJ, Schmitz MD, Strömberg CAE, Carlini AA, Ré GH, Crowley J (2013) A new chronology for middle Eocene–early Miocene South American land mammal ages. GSA Bulletin 125:539–555

Esteban G, Nasif N, Georgieff S (2014) Cronobioestratigrafía del Mioceno tardío–Plioceno temprano, Puerta de Corral Quemado y Villavil, provincia de Catamarca Argentina. Acta Geol Lilloana 26:165–188

Figueirido B, Janis CM, Pérez-claros JA et al (2012) Cenozoic climate change influences mammalian evolutionary dynamics. Proc Natl Acad Sci USA 109:722–727

Flynn JJ, Swisher CC III (1995) Cenozoic South American land mammal ages: correlation to global geochronologies. In: Berggren WA, Kent DV, Aubry M-P, Hardenbol J (eds) Geochronology, time-scales and global stratigraphic correlation: a unified framework for an historical geology. Soc Strat Geol Spec Pub, vol 54, pp 317–333

Flynn JJ, Wyss AR, Croft DA, Charrier R (2003) The Tinguiririca Fauna, Chile: biochronology, paleoecology, biogeography, and a new earliest Oligocene South American land mammal age. Palaeogeogr Palaeoclim Palaeoecol 195:229–259

Foote M (2000) Origination and extinction components of taxonomic diversity: Paleozoic and post-Paleozoic dynamics. Paleobiology 26:578–605

Forasiepi AM, Prevosti JF, Vera B, Turazzini GF, Echarri S, Garrido AC, Verzi D, Rasia LL, Schmidt GI, Esteban G, Krapovikas V (2014) The badlands from Mendoza and the Huayquerian Age: insights into the late Miocene. 4th International Palaeontologial Congress, Abstracts: 713

Goin FJ, Gelfo JN, Chornogubsky L, Woodburne MO, Martin T (2012) Origins, radiations, and distribution of South American mammals: from greenhouse to icehouse worlds. In:

Patterson BD, Costa LP (eds) Bones, clones, and biomes: an 80-million year history of Recent Neotropical mammals. University Chicago Press, Chicago, pp 20–50

Goin FJ, Woodburne MO, Martin GM, Chornogubsky L (2016) A brief history of South American metatherians: evolutionary contexts and intercontinental dispersals. Springer Earth System Sciences, Dordrecht

Isla F, Taglioretti M, Dondas A (2015) Revisión y nuevos aportes sobre la estratigrafía y sedimentología de los acantilados entre Mar de Cobo y Miramar, Provincia de Buenos Aires. RAGA 72:235–250

Kay RF, Madden RH, Vucetich MG, Carlini AA, Mazzoni MM, Re GH, Heizler M, Sandeman H (1999) Revised age of the Casamayoran South American Land Mammal 'Age'—climatic and biotic implications. Proc Nat Acad Sci 96:13235–13240

Kowalewski M, Flessa KW (1996) The fossil record of lingulide brachiopods and the nature of taphonomic mega biases. Geology 24:977–980

Kraglievich L (1934) La antigüedad Plioceno de las faunas de Monte Hermoso y Chapadmalal, deducidas de su comparación con las que le precedieron y sucedieron. Imprenta El Siglo Ilustrado, Montevideo

Maas MC, Anthony MRL, Gingerich PD, Gunnel GF, Krause DW (1995) Mammalian generic diversity and turnover in the late Paleocene and early Eocene of the Bighorn and Crazy Mountain Basins, Wyoming and Montana (USA). Palaeogeogr Palaeoclimatol Palaeoecol 115:181–207

Madden RH, Kay RF, Vucetich MG, Carlini AA (2010) Gran Barranca: a 23 million-year record of middle Cenozoic faunal evolution in Patagonia. In: Madden RH, Carlini AA, Vucetich MG, Kay RF (eds) The Paleontology of Gran Barranca: evolution and environmental change through the middle Cenozoic of Patagonia. Cambridge University of Press, New York

Manly BFJ (1997) Randomization bootstrap and Monte Carlo methods in biology. Chapman & Hall, London

Marshall LG, Cifelli RL (1990) Analysis of changing diversity patterns in Cenozoic land mammal age faunas, South America. Palaeovertebrata 19:169–210

Marshall LG, Patterson B (1981) Geology and geochronology of the mammal-bearing Tertiary of the valle de Santa María and río Corral Quemado, Catamarca province, Argentina. Fieldiana Geology 9:1–80

Marshall LG, Pascual R, Curtis GH, Drake RE (1977) South American geochronology: radiometric time scale for middle to late Tertiary mammal-bearing horizons in Patagonia. Science 195:1325–1328

Marshall LG, Butler RF, Drake RE, Curtis DH (1981) Calibration of the beginning of the age of mammals in Patagonia. Science 212:43–45

Marshall LG, Webb SD, Sepkoski JJ Jr, Raup DM (1982) Mammalian evolution and the great American interchange. Science 215:1351–1357

Marshall L, Hoffstetter R, Pascual R (1983) Mammals and stratigraphy: geochronology of the continental mammal-bearing Tertiary of South America. Palaeovertebrata, Mémoire Extraordinaire, pp 1–93

Marshall L, Berta A, Hoffstetter R, Pascual R, Reig O, Bombin M, Mones A (1984) Mammals and stratigraphy: geochronology of the continental mammal-bearing Quaternary of South America. Palaeovertebrata, Mémoire Extraordinaire, pp 1–76

Marshall LG, de Muizon C, Sigé B (1985) The "Rosetta stone" for mammalian evolution in South America. Nat Geogr Res 1:274–288

Noto CR (2011) Hierarchical control of terrestrial vertebrate taphonomy over space and time: discussion of mechanisms and implications for vertebrate paleobiology. In: Allison PA, Bottjer DJ (eds) Taphonomy: process and bias through time. Topics in Geobiology, vol 32, pp 287–336

Palombo MR, Alberdi MT, Azanza B et al (2009) How did environmental disturbances affect carnivoran diversity? A case study of the Plio-Pleistocene Carnivora of the North-Western Mediterranean. Evol Ecol 23:569–589

Pascual R (2006) Evolution and geography: the biogeographic history of South American land mammals. Ann Missouri Bot Gard 93:209–230

Pascual R, Odreman Rivas O (1971) Evolución de las comunidades de los vertebrados del Terciario argentino, los aspectos paleozoogeográficos y paleoclimáticos relacionados. Ameghiniana 8:372–412

Pascual R, Ortiz-Jaureguizar E (1990) Evolving climates and mammal faunas in Cenozoic South America. J Human Evol 19:23–60

Patterson B, Pascual R (1972) The fossil mammal fauna of South America. In: Keast A, Erk FC, Glass B (eds) Evolution of mammals and Southern continents. State Univesity of New York Press, Albany

Pérez LM (2013) Nuevo aporte al conocimiento de la edad de la Formación Paraná, Mioceno de la Provincia de Entre Ríos, Argentina. In: Brandoni D, Noriega JI (eds) El Neógeno de la Mesopotamia Argentina, Publicación especial APA, vol 14, pp 7–12

Perkins ME, Fleagle JG, Heizler MT, Nash B, Bown TM, Tauber AA, Dozo MT (2012) Tephrochronology of the Miocene Santa Cruz and Pinturas formations, Argentina. In: Vizcaíno SF, Kay RF, Bargo MS (eds) Early Miocene paleobiology in Patagonia. Cambridge University Press, Cambridge, pp 23–40

Prevosti FJ (2006) New materials of Pleistocene cats (Carnivora, Felidae) from southern South America, with comments on biogeography and the fossil record. Geobios 39:679–694

Prevosti FJ, Pardiñas UFJ (2009) Comment on "The oldest South American Cricetidae (Rodentia) and Mustelidae (Carnivora): late Miocene faunal turnover in central Argentina and the Great American Biotic Interchange" by Verzi DH, Montalvo CI. Palaeogeogr Palaeoclimatol Palaeoecol 280:543–547

Prevosti FJ, Pomi L (2007) Revisión Sistemática de *Smilodontidion riggii*. Argent. Cienc. Nat., n. s. 9:67–77

Prevosti F, Soibelzon L (2012) Evolution of the South American carnivores (Mammalia, Carnivora): a paleontological perspective. In: Patterson BD, Costa LP (eds) Bones, clones, and biomes: an 80-million year history of Recent Neotropical mammals. University Chicago Press, Chicago, pp 102–122

Prevosti FJ, Gasparini GM, Bond M (2006) Systematic position of a specimen previously assigned to carnivora from the Pliocene of Argentina and its implications for the great American biotic interchange. Neues Jahrb Geol Paläontol Abh 242:133–144

Prevosti FJ, Tonni EP, Bidegain JC (2009) Stratigraphic range of the large canids (Carnivora, Canidae) in South America, and its relevance to quaternary biostratigraphy. Quat Int 210:76–81

Prevosti FJ, Forasiepi A, Zimicz N (2013) The evolution of the Cenozoic terrestrial mammalian predator guild in South America: competition or replacement? J Mamm Evol 20:3–21

Reguero MA, Candela AM (2011) Late Cenozoic mammals from the northwest of Argentina. In: Salfity R, Marquillas MR (eds) Cenozoic geology of the Central Andes of Argentina. SCS Publishers, Salta, pp 411–426

Rico Y, Bidegain JC (2013) Magnetostratigraphy and environmental magnetism in a sedimentary sequence of Miramar, Buenos Aires, Argentina. Quat Int 317:53–63

Riggs ES, Patterson B (1939) Stratigraphy of late miocene and pliocene deposits of the province of Catamarca (Argentina) with notes on the faune. Physis 14:143–162

Rovereto C (1914) Los estratos araucanos y sus fósiles. Anal Mus Nac Hist Nat Buenos Aires 25:1–247

Schultz P, Zarate M, Hames W et al (1998) A 3.3-Ma impact in Argentina and possible consequences. Science 282:2061–2063

Schultz PH, Rate M, Hames WE et al (2006) The record of Miocene impacts in the Argentine Pampas. Meteorit Planet Sci 41:749–771

Simpson GG (1940) Review of the mammal-bearing Tertiary of South America. Proc Am Phil Soc 83:649–709

Simpson GG (1950) History of the fauna of Latin America. Am Sci 38:261–389

Simpson GG (1980) Splendid isolation: the curious history of South American mammals. Yale University Press, New Haven & London

Stehli FG, Webb SD (1985) The great American biotic interchange. Plenum Press, New York

Tambussi C, Degrange F (2011) South American and Antarctic continental Cenozoic birds. Paleobiogeographic affinities and disparities, Springer Earth System Sciences, Dordrecht

Tejedor MF, Goin FJ, Gelfo JN, López G, Bond M, Carlini AA, Scillato-Yané GJ, Woodburne MO, Chornogubsky L, Aragón E, Reguero M, Czaplewski N, Vincon S, Martin GM, Ciancio M (2009) New early Eocene mammalian fauna from Western Patagonia, Argentina. Amer Mus Novit 3638:1–42

Tomassini RL, Montalvo CI, Deschamps CM, Manera T (2013) Biostratigraphy and biochronology of the Monte Hermoso Formation (early Pliocene) at its type locality, Buenos Aires province, Argentina. J South Am Earth Sci 48:31–42

Tonni E (2009) Los mamíferos del Cuaternario de la Región Pampeana de Buenos Aires, Argentina. In: Ribeiro AM, Bauermann SG, Cherer CS (eds) Quaternario Do Rio Grande do Sul: integrando conocimientos. Monografias da Sociedade Brasileira de Paleontologia, Porto Alegre, pp 193–205

Tonni EP, Scillato Yané GJ, Cione AL, Carlini AA (1998) Bioestratigrafía del Mioceno continental en el curso inferior del arroyo Chasicó, provincia de Buenos Aires. In: Resúmenes del VII Congreso Argentino de Paleontología y Bioestratigrafía, p 135

Verzi DH, Montalvo CI (2008) The oldest South American Cricetidae (Rodentia) and Mustelidae (Carnivora): late miocene faunal turnover in central Argentina and the great American biotic interchange. Palaeogeogr Palaeoclimatol Palaeoecol 267:284–291

Verzi DH, Montalvo CI, Deschamps CM (2008) Biostratigraphy and biochronology of the late miocene of central Argentina: evidence from rodents and taphonomy. Geobios 41:145–155

Vizcaíno SF, Kay RF, Bargo MS (2012) Early Miocene Paleobiology in Patagonia. Cambridge University Press, Cambridge

Webb SD (1991) Ecogeography and the great American interchange. Paleobiology 17:266–280

Webb SD (2006) The great American biotic interchange: patterns and processes. Ann Mo Bot Gard 93:245–257

Wilf P, Cúneo R, Escapa IH, Pol D, Woodburne MO (2013) Splendid and seldom isolated: the paleobiogeography of Patagonia. Annu Rev Earth Planet Sci 41:561–603

Woodburne MO (2010) The great American biotic interchange; dispersals, tectonics, climate, sea level and holding pens. J Mamm Evol 17:245–264

Woodburne MO, Cione AL, Tonni EP (2006) Central American provincialism and the great American biotic interchange. In: Carranza-Castañeda O, Lindsay EH (eds) Advances in late Tertiary vertebrate paleontology in Mexico and the great American biotic interchange, vol 4. Inst de Geol y Centro de Geociencias. UNAM, Publ Esp, pp 73–101

Woodburne MO, Goin FJ, Bond M, Carlini AA, Gelfo JN, López GM, Iglesias A et al (2014a) Paleogene land mammal faunas of South America: a response to global climatic changes and indigenous floral diversity. J Mammal Evol 21:1–73

Woodburne MO, Goin FJ, Raigemborn MS, Heizler M, Gelfo JN, Oliveira EV (2014b) Revised timing of the South American early Paleogene land mammal ages. J S Amer Earth Sci 54:109–119

Yrigoyen MR (1994) Revisión estratigráfica del Neógeno de las Huayquerías de Mendoza septentrional, Argentina. Ameghiniana 31:125–138

Zamorano M, Taglioretti M, Zurita AE, Scillato-Yané GJ, Scaglia YF (2014) El registro más antiguo de *Panochthus* (Xenarthra, Glyptodontidae). Estud Geol 70:1–6

Zimicz N (2014) Avoiding competition: the ecological history of late Cenozoic metatherian carnivores in South America. J Mamm Evol 21:383–393

Chapter 6
Evolution and Biological Context of South American Mammalian Carnivores During the Cenozoic and the Biological Context

Abstract The process by which successive groups using the same resources occupy the same geographic area through time is frequently attributed to competition. Several authors have argued that competitive displacement was the cause of the decline and extinction of Sparassodonta, due to the introduction of carnivorans into South America about 8–7 Ma, although this view has been recently criticized. The diversity of Sparassodonta was low relative to that of Carnivora throughout the Cenozoic. The greatest peak in sparassodontan diversity was during the early Miocene (Santacrucian), with 11 species. After the late Miocene (Huayquerian), sparassodont diversity decreased and the group became extinct in the mid-Pliocene (~3 Ma, Chapadmalalan). In the late Miocene–mid Pliocene (Huayquerian–Chapadmalalan), the fossil record shows that sparassodonts and carnivorans overlapped. During this time, carnivoran diversity consisted of four or fewer species; thereafter, it expanded to more than 20 species in the early–Middle Pleistocene (Ensenadan). Initially, Carnivora was represented by middle-sized, omnivorous species, with large omnivores first represented in the mid-Pliocene (Chapadmalalan). By contrast, over this period, Sparassodonta was represented by both large and small hypercarnivores and a single large omnivorous species. We review hypotheses of replacement using the available information and perform new analyses to test the effect of sampling bias, ecological overlap between clades, and the relevance of environmental and faunistic changes for the evolution of sparassodonts. From this review of the fossil record, it is suggested that stochastic mechanisms other than competitive displacement may have caused the decline and extinction of Sparassodonta, possibly as part of a larger faunistic turnover related to multicausal biological and physical factors. Similarly, at the Pleistocene/Holocene boundary, an extinction event affected large mammals in South America, including large carnivorans, in the context of a multicausal event that involved human presence as well as collateral factors.

Keywords Sparassodonta · Carnivora · Competitive displacement
Ecological replacement · Faunistic turnover · Multicausal event

© Springer International Publishing AG 2018
F.J. Prevosti and A.M. Forasiepi, *Evolution of South American Mammalian Predators During the Cenozoic: Paleobiogeographic and Paleoenvironmental Contingencies*, Springer Geology, https://doi.org/10.1007/978-3-319-03701-1_6

6.1   Introduction

The ecological interaction between species is one of the drivers of macroevolutionary processes. Using the analogy of the Red Queen in *Through the Looking Glass*, Van Valen (1973) proposed that organisms are in a constant evolutionary battle to out-compete competitors. In this context, ecological competition refers to the situation in which a local population of one species reduces the rate of expansion of another (Sepkoski 2001). Competitive displacement (competitive exclusion or active displacement, sensu Krause 1986) in the fossil record concerns contexts in which one taxon wanes while other waxes—the so-called "double wedge" pattern (Sepkoski 2001)—with the consequent extinction of one of the species involved. Relevant studies include classical ecological pairs such as gastropods and brachiopods (Gould and Calloway 1980; Sepkoski 2001), basal archosauromorphs and dinosaurs (Brusatte et al. 2008; Langer et al. 2009), multituberculates and rodents (Krause 1986), creodonts and carnivorans (Van Valkenburgh 1999; Friscia and Van Valkenburgh 2010), and different clades of carnivorans (Silvestro et al. 2015).

In South America, the decline and extinction of the native mammalian predators (Sparassodonta) was classically interpreted in relation to the arrival of placental carnivorans (Carnivora), during the late Miocene–Pliocene (e.g., Simpson 1950, 1969, 1971, 1980; Patterson and Pascual 1972; Savage 1977; Werdelin 1987; Wang et al. 2008). However, this hypothesis has been questioned and even rejected (e.g., Marshall 1977, 1978; Reig 1981; Bond 1986; Pascual and Bond 1986; Goin 1989, 1995; Ortiz Jaureguizar 1989, 2001; Marshall and Cifelli 1990; Alberdi et al. 1995; Forasiepi et al. 2007; Forasiepi 2009; Prevosti et al. 2013; Zimicz 2014; López-Aguirre et al. 2017).

Different models have been applied to try to understand the evolution of two convergent clades occupying the same geographic area at the same time. The simplest model to predict the possibility of competitive displacement is linear decrease in diversity and/or abundance of one taxon, associated with increase in diversity and/or abundance in another (Benton 1983; Krause 1986; Van Valkenburgh 1999). Other predictions have coupled logistic functions describing local population sizes of competing species in local environments (Sepkoski 1996, 2001; Sepkoski et al. 2000). Under this model, both groups increase their diversity/ abundance until a threshold is reached; subsequently, the out-competed taxon declines. Another model implies that both groups persist for a long time until an external perturbation occurs, when the diversity/abundance of one of them decreases in favor of the other ("incumbent replacement" sensu Rosenzweig and McCord 1991). Recently, new approaches have been made with Bayesian analyses (e.g., Silvestro et al. 2015; Pires et al. 2015).

Similarly, environmental changes have been largely discussed in relation to the extinction processes of different mammalian groups. In examples that have been studied, such as the "Grande Coupure" at the Eocene/Oligocene boundary in Eurasia, not one but several taxonomic groups are affected, triggering a faunistic turnover. In South America, this complex has been identified in relation to the cooling of Patagonia also at the Eocene/Oligocene boundary (Goin et al. 2010, 2012), or Andean orogeny and Southern cone aridification by the mid-Miocene (Pascual and Ortiz Jaureguizar 1990) (Chap. 2). In a similar context, our last revision (Prevosti et al. 2013) suggested that the extinction of Sparassodonta was related to environmental and faunistic changes occurring between the middle Miocene and the Pliocene. In this chapter, we review the pattern of evolution of carnivores in SA and evaluate the hypothesis of competitive displacement and the role of environmental factors in faunal turnover.

6.2   Changes in Predator Diversity: The Last Sparassodonts and First Carnivorans in South America

Our revised dataset produced diversity curves very similar to ones presented earlier (Prevosti et al. 2013; see also Chaps. 3–5; Fig. 6.1). During the earliest 20 million years of the Cenozoic, until the middle Eocene (Casamayoran), sparassodont taxonomic diversity and ecological disparity were low. It is worth mentioning that the phylogenetic position of the early Paleocene (Tiupampan) metatherians as sparassodonts is still inconclusive (Chap. 3); however, as a means of expressing the idea that the group was indeed present (e.g., de Muizon 1994, 1998; de Muizon et al. 1997, 2015), we include one Tiupampan metatherian for our analyses. The subsequent Peligran Age has provided only one isolated astragalus that resembles in size and shape of a fox-like borhyaenid (Goin et al. 2002). Unquestioned sparassodonts were present by the early Eocene (Itaboraian Age), as demonstrated by *Patene simpsoni*. The first peak in diversity is recorded by the middle Eocene (Casamayoran) with nine species, but this number may be biased due to averaging of faunal associations (Chap. 5). No or little record exists for the Mustersan–Tinguirirican, Friasian–Colloncuran, or Mayoan. The major peak in sparassodont diversity is recorded in the Santacrucian with 11 species (Fig. 6.2), followed by a decrease after the Huayquerian (Fig. 6.3) until their last records in the Chapadmalalan. Carnivoran diversity is low during the late Miocene–Pliocene, with a notable increase in the early–Middle Pleistocene (Ensenadan) (Fig. 6.4). The addition of Lazarus taxa to the analysis helps to reduce the effects of apparent declines and missing records, but it does not completely eliminate biases (Chap. 5).

   Seven late Miocene (Huayquerian Age; Fig. 6.3) sparassodont species were included in our analysis: *Stylocynus paranensis*, *Borhyaenidium altiplanicus*,

Fig. 6.3 Artist's portrayal of a landscape during the late Miocene (Huayquerian Age) in Pampa. In the foreground, left to right the carnivores *Cyonasua* (Carnivora) and *Thylacosmilus* (Sparassodonta). The center of the image is dominated by *Argentavis* (Theratornithidae), the largest extinct flying bird, with a ca 7-m wingspan. In the distant left background, three hoplophorine glyptodonts (Glyptodontidae) and a group of *Xotodon* (Notoungulata). Artist Jorge Blanco

Fig. 6.4 Artist's portrayal of a landscape during the Pleistocene (Ensenadan Age) in Pampa. *Smilodon* (Carnivora) crouches near a group of *Antifer* (Cervidae). In the background, one *Glyptodon* (Glyptodontidae) and a group of *Stegomastodon* (Gomphotheriidae). Artist Jorge Blanco

Borhyaenidium musteloides, Notictis ortizi, and *Thylacosmilus atrox,* as well as *Eutemnodus americanus* (MACN-A 4975, 3990, 3991) and *Borhyaena* sp. (MACN-PV 13207), represented by fragmentary material. The alpha taxonomy of the last two taxa remains uncertain, but they nonetheless record the presence of at least one more large-sized hypercarnivorous taxon in the late Miocene of southern SA.

Late Miocene–early Pliocene carnivorans were represented by procyonids (*Cyonasua* and *Chapalmalania*). Other lineages (mustelids, foxes) were represented only since the Vorohuean (Prevosti and Soibelzon 2012; Prevosti et al. 2013). All living groups (as well as certain fossil ones) have known Pleistocene representatives. As was discussed in Chap. 4, the presence of felids (i.e., *"Felis" pumoides, Smilodontidion riggii*) and a skunk (i.e., *Conepatus altiramus*) is not supported for the Pliocene, but in some of our analyses, we explore the potential impact of these dubious Pliocene records for our interpretations.

6.3 Diet, Body Size, and the Evolution of South American Carnivore Faunas

In view of the estimates based on RGA index and body masses, we consider that sparassodonts were mostly hypercarnivores, with the curve of hypercarnivory mostly corresponding to taxonomic diversity (Tables 6.1, 6.2, 6.3 and 6.4). There is only one mesocarnivore taxon in the Deseadan (*Pharsophorus tenax*), while omnivores have low diversity throughout and are mostly restricted to the Paleogene, with the exception of the Huayquerian and Laventan (*Stylocynus paranensis* and *Hondadelphys fieldsi,* respectively; Table 6.1). Body mass estimates indicate that middle-sized sparassodonts were few and restricted to the Mustersan, Colhuehuapian, Laventan, and Chasicoan (Table 6.1). The opposing extremes, small and large sparassodonts, were common, and their abundance follows the diversity curve. Body mass and RGA values together indicate that sparassodonts were small and omnivorous until the Casamayoran, when large body sizes and hypercarnivory were also the most frequent categories. The largest body size (~ 200 kg, *Proborhyaena gigantea*) was recorded for the Deseadan, but other large taxa (~ 100 kg, *Thylacosmilus*) managed to persist until the late Miocene–Pliocene. Median body size reached ca. 20 kg in the Deseadan and remained near or above this value until the extinction of the group. Median size dropped to ca. 6 kg in the Santacrucian when a high diversity of small taxa was present (Prevosti et al. 2012; Ercoli et al. 2014), but recovered to ca. 13 kg in the Chasicoan (Table 6.3). Relative grinding area (RGA) index range is greatest in the Casamayoran (range ~ 0.7–0), followed by the Huayquerian (range ~ 0.6–0); while it is ~ 0.3 for most other ages (Table 6.3). In the Huayquerian, middle-sized hypercarnivores were not in evidence but small and large hypercarnivores were present; no mesocarnivore has been recorded, and there was only a single large

Table 6.1 Abundance of each body size and diet class through the Cenozoic in South America

Age	Spa Large	Spa Med	Spa Small	Spa Hyper	Spa Meso	Spa Omni	Car Large	Car Med	Car Small	Car Hyper	Car Meso	Car Omni
Tiupampan	0	0	1	0	0	1	0	0	0	0	0	0
Peligran	0	0	0	0	0	0	0	0	0	0	0	0
Itaboraian	0	0	1	0	0	1	0	0	0	0	0	0
Riochican	0	0	0	0	0	0	0	0	0	0	0	0
Casamayoran	5	0	4	7	0	2	0	0	0	0	0	0
Mustersan	1	1	1	2	0	1	0	0	0	0	0	0
Tinguirirican	0	0	0	0	0	0	0	0	0	0	0	0
Deseadan	6	0	2	7	1	0	0	0	0	0	0	0
Colhuehuapian	4	1	2	7	0	0	0	0	0	0	0	0
Santacrucian	5	0	6	11	0	0	0	0	0	0	0	0
Friasian	2	0	2	4	0	0	0	0	0	0	0	0
Colloncuran	2	0	1	3	0	0	0	0	0	0	0	0
Laventan	3	1	1	4	0	1	0	0	0	0	0	0
Mayoan	0	0	0	0	0	0	0	0	0	0	0	0
Chasicoan	1	3	0	4	0	0	0	0	0	0	0	0
Huayquerian	4	0	3	6	0	1	0	4	0	0	0	4
Montehermosan	1	0	1	2	0	0	0	1	0	0	0	1
Chapadmalalan	1	0	1	2	0	0	2	1	0	0	0	3
Barrancalobian	0	0	0	0	0	0	0	0	0	0	0	0
Vorohuean	0	0	0	0	0	0	1	0	2	1	0	2
Sanandresian	0	0	0	0	0	0	0	0	2	1	0	1
Ensenadan	0	0	0	0	0	0	10	2	10	15	1	7
Bonaerian	0	0	0	0	0	0	0	0	1	1	0	0
Lujanian	0	0	0	0	0	0	11	5	20	16	4	16
Platan	0	0	0	0	0	0	4	7	12	7	4	12

Spa Sparassodonta; *Car* carnivora; *Large* large body size class; *Med* medium body size class; *Small* small body size class; *Hyper* hypercarnivores; *Meso* mesocarnivores; *Omni* omnivores

Table 6.2 Abundance of each predator body size and diet class through the Cenozoic in South America considering Lazarus taxa

Age	Spa Large	Spa Med	Spa Small	Spa Hyper	Spa Meso	Spa Omni	Car Large	Car Med	Car Small	Car Hyper	Car Meso	Car Omni
Tiupampan	0	0	1	0	0	1	0	0	0	0	0	0
Peligran	0	0	1	0	0	1	0	0	0	0	0	0
Itaboraian	0	0	1	0	0	1	0	0	0	0	0	0
Riochican	0	0	1	0	0	1	0	0	0	0	0	0
Casamayoran	5	0	4	7	0	2	0	0	0	0	0	0
Mustersan	1	1	1	2	0	1	0	0	0	0	0	0
Tinguirirican	0	0	0	0	0	0	0	0	0	0	0	0
Deseadan	6	0	2	7	1	0	0	0	0	0	0	0
Colhuehuapian	4	1	2	7	0	0	0	0	0	0	0	0
Santacrucian	5	0	6	11	0	0	0	0	0	0	0	0
Friasian	3	1	3	5	0	0	0	0	0	0	0	0
Colloncuran	2	1	1	4	0	0	0	0	0	0	0	0
Laventan	3	1	1	4	0	1	0	0	0	0	0	0
Mayoan	1	0	0	1	0	0	0	0	0	0	0	0
Chasicoan	1	3	0	4	0	0	0	0	0	0	0	0
Huayquerian	4	0	3	6	0	1	0	4	0	0	0	4
Montehermosan	1	0	2	3	0	0	0	1	0	0	0	1
Chapadmalalan	1	0	1	2	0	0	2	1	0	0	0	3
Barrancalobian	0	0	0	0	0	0	1	1	0	0	0	2
Vorohuean	0	0	0	0	0	0	1	1	2	1	0	3
Sanandresian	0	0	0	0	0	0	0	1	3	1	0	3
Ensenadan	0	0	0	0	0	0	10	2	10	15	1	7
Bonaerian	0	0	0	0	0	0	7	1	6	6	1	7
Lujanian	0	0	0	0	0	0	11	5	20	16	4	16
Platan	0	0	0	0	0	0	5	8	16	11	4	14

Spa Sparassodonta; *Car* carnivora; *Large* large body size class; *Small* small body size class; *Med* medium body size class; *Hyper* hypercarnivores; *Meso* mesocarnivores; *Omni* omnivores

Table 6.3 Mammalian predator body mass distribution through the Cenozoic in South America

Age	Med S	Min S	Max S	Range S	Med C	Min C	Max C	Range C	Med S+L	Min S+L	Max S +L	Range S+L	Med C+L	Min C+L	Max C +L	Range C+L
Tiupampan																
Peligran																
Itaboraian	1.35	1.35	1.35	0.00					1.35	1.35	1.35	0.00				
Riochican	1.35	1.35	1.35	0.00					1.35	1.35	1.35	0.00				
Casamayoran	18.50	1.35	31.50	30.15					18.50	1.35	31.50	30.15				
Mustersan	8.90	1.00	38.50	37.50					8.90	1.00	38.50	37.50				
Tinguiririran																
Deseadan	29.40	2.06	200.00	197.94					29.40	2.06	200.00	197.94				
Colhuehuapian	16.85	1.70	40.00	38.30					16.85	1.70	40.00	38.30				
Santacrucian	6.60	1.17	43.87	42.70					6.60	1.17	43.87	42.70				
Friasian	19.30	2.11	36.40	34.29					19.30	2.11	36.40	34.29				
Colloncuran	16.00	0.89	22.00	21.11					22.00	0.89	16.00	15.11				
Laventan	17.00	3.70	48.53	44.83					17.00	3.70	48.53	44.83				
Mayoan																
Chasicoan	13.55	8.27	19.20	10.93					13.55	8.27	19.20	10.93				
Huayquerian	31.05	1.17	117.40	116.23	13.32	11.37	14.73	3.36	31.05	1.17	117.40	116.23	13.32	11.37	14.73	3.36
Montehermosan	59.94	2.48	117.40	114.92	15.32	15.32	15.32	0.00	59.49	1.57	117.40	114.92	15.32	15.32	15.32	0.00
Chapadmalalan	59.69	1.98	117.40	115.42	63.51	14.26	88.13	73.87	59.69	1.98	117.40	115.42	63.51	14.26	88.13	73.87
Barrancalobian													50.81	13.49	88.13	74.64
Vorohuean					31.19	2.12	88.13	86.01					8.41	2.12	88.13	86.01
Sanandresian					3.16	2.12	4.20	2.08					3.77	2.12	13.49	2.08
Ensenadan					80.00	0.23	900.00	899.77					80.00	0.23	900.00	899.77
Bonaerian					243.22	2.36	600.00	597.64					7.75	2.36	600.00	597.64
Lujanian					63.78	0.23	600.00	599.77					63.78	0.23	600.00	599.77
Platan					19.73	0.12	175.00	174.88					5.00	0.12	175.00	174.88

S Sparassodonta; C carnivora; Med median; Min minimum; Max Maximum; Range body mass range; +L Lazarus taxa included in the calculation

Table 6.4 Relative grinding area of lower carnassial through the Cenozoic in South America

Age	Med S	Min S	Max S	Range S	Med C	Min C	Max C	Range C	Med S +L	Min S +L	Max S +L	Range S +L	Med C +L	Min C +L	Max C +L	Range C +L
Tiupampan																
Peligran																
Itaboraian	0.70	0.70	0.70	0.00					0.7	0.7	0.7	0				
Riochican												0				
Casamayoran	0.17	0.00	0.70	0.70					0.17	0	0.7	0.7				
Mustersan	0.17	0.17	0.70	0.53					0.17	0.17	0.7	0.53				
Tinguirirican																
Deseadan	0.17	0.00	0.53	0.53					0.17	0	0.53	0.53				
Colhuehuapian	0.17	0.00	0.33	0.33					0.17	0	0.33	0.33				
Santacrucian	0.27	0.00	0.34	0.34					0.27	0	0.34	0.34				
Friasian	0.17	0.00	0.33	0.33					0.17	0	0.33	0.33				
Colloncuran	0.30	0.00	0.30	0.30					0.3	0	0.3	0.3				
Laventan	0.30	0.00	0.63	0.63					0.3	0	0.63	0.63				
Mayoan																
Chasicoan	0.27	0.26	0.30	0.04					0.265	0.26	0.3	0.04				
Huayquerian	0.30	0.00	0.61	0.61	0.95	0.88	0.91	0.03	0.3	0	0.61	0.61	0.91	0.88	0.95	0.07
Montehermosan	0.17	0.00	0.33	0.33	0.86	0.86	0.86	0.00	0.31	0	0.33	0.33	0.86	0.86	0.86	0.00
Chapadmalalan	0.16	0.00	0.32	0.32	0.79	0.75	0.77	0.02	0.16	0	0.32	0.32	0.77	0.75	0.79	0.04
Barrancalobian													0.86	0.77	0.95	0.18
Vorohuean					0.55	0.55	0.79	0.24					0.67	0.55	0.79	0.24
Sanandresian					0.46	0.35	0.56	0.21					0.56	0.35	0.95	0.60
Ensenadan					0.40	0.00	1.22	1.22					0.40	0	1.22	1.22
Bonaerian					0.00	0.00	0.00	0.00					0.54	0	1.22	1.22
Lujanian					0.50	0.00	1.27	1.27					0.50	0	1.27	1.27
Platan					0.52	0.00	2.08	2.08					0.50	0	2.08	2.08

S Sparassodonta; *C* carnivora; *Med* median; *Min* minimum; *Max* maximum; *Range* body mass range; *+L* Lazarus taxa included in the calculation

omnivore. In the Montehermosan and Chapadmalalan, only hypercarnivores in the small and large size ranges remained (Table 6.4).

During the late Miocene–Pliocene, the ecological disparity of SA carnivorans was limited (Prevosti and Soibelzon 2012; Prevosti et al. 2013; Chap. 4). Broad disparity in body size (0.23–900 kg) and diet (RGA between 0 and >1) existed from the Ensenadan onward (Tables 6.1 and 6.2; the drop observed in the Bonaerian is due to fossil record bias; see Chap. 5). New body size estimates classify *Cyonasua* spp. as a middle-sized omnivore (contra small omnivore in Prevosti and Soibelzon 2012; Prevosti et al. 2013). Large omnivores (*Chapalmalania* spp.) were present during the Chapadmalalan, and small omnivores and hypercarnivores (*Lycalopex cultridens* and *Galictis sorgentinii*, respectively) in the Vorohuean (Tables 6.1 and 6.2, Chap. 4). After the Lujanian, large hypercarnivores became fewer as a consequence of the Late Pleistocene–Holocene extinctions (see below).

As in the case of diversity, biases in the fossil record (Chap. 5) also affect the distribution of ecological types. The inclusion of Lazarus taxa does not modify the general pattern, but it certainly affects interpretation of the carnivoran fossil record for the Bonaerian, making it more similar to the Lujanian and Ensenadan (Table 6.2).

To compare and explore ecological variation within each age, we performed a principal component analysis on the correlation matrix, with body size and diet classes analyzed per age (Fig. 6.5). The first axis organizes taxonomically more diverse faunas (Lujanian, Ensenadan, Platan) to the right and less diverse ones (e.g., Bonaerian, Marplatan Subages) to the left. The second axis orders the faunas by the relative number of large hypercarnivores (positively correlated with axis 2 scores) and mesocarnivores, omnivores, and medium-sized taxa (negatively correlated with axis 2 scores). Ensenadan, Santacrucian, and Casamayoran Ages have the highest scores on this axis, while Chasicoan, Huayquerian, and Platan are at the opposite extreme (Fig. 6.5). The position of certain intervals (e.g., Bonaerian, Itaboraian, and Marplatan Subages), lying at the negative extreme of axis 1 and the middle of axis 2, is clearly conditioned by their low diversity, something that could be partially explained by biases in the fossil record. This is clear for the Bonaerian, because the inclusion of Lazarus taxa generates a similar morphospace, but this unit is positive for axis 1 and axis 2 and thus more similar to the Lujanian and Ensenadan. Other ages with low diversities do not change significantly when Lazarus taxa are included, and this suggests that the low taxonomic diversity and ecological diversity of some intervals (e.g., Marplatan Subages and the Chapadmalalan in particular) is a real pattern. Other methods (e.g., ghost lineages, Bayesian, capture–recapture; e.g., Liow and Finarelli 2014; Silvestro et al. 2015; Pires et al. 2015; Finarelli and Liow 2016) should be considered to further evaluate diversity and ecological characterization.

Finally, we explored the ecological "disparity" (i.e., the area occupied by a morphospace, as well as, the morphospace area divided by the number of taxa) of well-sampled ages (Table 6.5) using a morphospace generated by body size and RGA (Fig. 6.6), and measuring spatial distribution using the software Past 2.17c (nearest neighbor distance with Donnelly edge correction; Clark and Evans 1954;

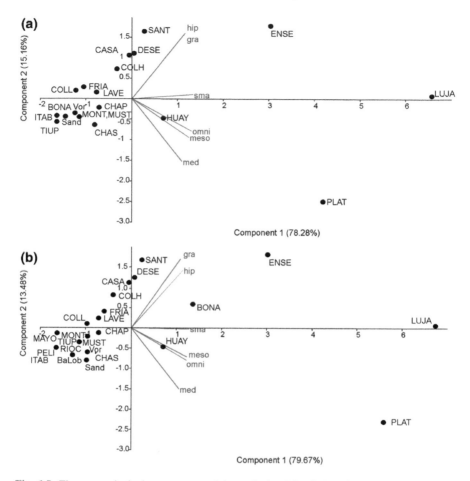

Fig. 6.5 First two principal components of the analysis of South American Ages, based on the abundance of each body size and diet class excluding **a** or excluding or including **b** Lazarus Taxa. *TIUP* Tiupampan; *PELI* Peligran; *ITAB* Itaboraian; *RIOC* Riochican; *CASA* Casamayoran; *MUST* Mustersan; *DESE* Deseadan; *COLH* Colhuehuapian; *SANT* Santacrucian; *FRIA* Friasian; *COLL* Colloncuran; *LAVE* Laventan; *MAYO* Mayoan; *CHAS* Chasicoan; *HUAY* Huayquerian; *MONT* Montehermosan; *CHAP* Chapadmalalan; *BaLob* Barrancalobian; *VOR* Vorohuean; *Sand* Sanandresian; *ENSE* Ensenadan; *BONA* Bonaerian; *LUJA* Lujanian; *PLAT* Platan; *gra* large size; *med* middle size; *sma* small size; *hip* hypercarnivores; *meso* mesocarnivores; *omni* omnivores. Green lines represent the loading of each ecological variable

Hammer et al. 2001; Hammer 2016). We judged disparity according to the density of the distribution of taxa. Our first measurement evaluates whether the taxa are distributed in clustered, random, or ordered (or over-dispersed) pattern. Unlike studies involving extant faunas, our analysis includes faunas at a continental scale, over intervals of a few thousands to millions of years. Averaging may generate false clustering if different taxa with similar ecological types are included. On the other

Table 6.5 Nearest neighbor distance (R) analysis of several South American Ages

Age	R	p	Area	Area/Taxa
Casamayoran	2.3573	<0.000000001	7.741	0.860
Deseadan	5.1261	<0.000000001	50.220	6.277
Santacrucian	2.4898	<0.00000001	6.817	0.620
Huayquerian	3.6299	<0.00000001	52.715	5.271
Chapadmalalan	2.1828	0.0000004	57.471	11.494
Chapadmalalan*	3.39	<0.00000001	86.957	10.870
Ensenadan	5.0943	<0.000000001	629.327	28.606
Lujanian	2.5816	<0.0000000001	478,424	13.290
Platan	2.8744	<0.000000001	206.297	8.970

p Probability that observed R comes from a random Poisson distribution; *Area* refers to the morphospace covered by each fauna; *Area/taxa* is the morphospace covered by each fauna divided by the number of taxa in each fauna. R below 1 indicates a clustered pattern of distribution, while above 1 indicates an ordered (or over-dispersed) pattern. Chapadmalalan* includes the doubtful Pliocene carnivorans (i.e., *Conepatus altiramus*, "*Felis*" *pumoides*, *Smilodontidion riggii*). The following body mass (BM) and relative grinding area (RGA) of lower carnassial were assigned as follows. *Conepatus altiramus*: BM, 2 kg and RGA, 1.1; "*Felis*" *pumoides*: BM, 50 kg and RGA, 0; and *Smilodontidion riggii*: BM, 150 kg and RGA, 0

hand, bias in fossil preservation could mask spatial distribution patterns, something that we try to minimize by analyzing only well-sampled faunas (see Chap. 5).

Unexpectedly, and because of data averaging, all the analyzed faunas showed significant ordered (or over-dispersed) distribution patterns (nearest neighbor distance well above 1) congruent with a well-structured guild pattern in which separation in body size and diet (RGA) acts to minimize competition between taxa. Ercoli et al. (2014) found the same pattern for the Santa Cruz Fm. (Santacrucian) using a similar approach, but including also locomotor habits, which minimized predator overlap. Unfortunately, the data are insufficient to include this ecological aspect in our analyses. The inclusion of inferences concerning carnivoran locomotion in Pleistocene and extant taxa may minimize overlap (Chap. 4).

Ecological disparity (Table 6.5) increases in the Deseadan due to a significant increase in the maximum body size, and in the Huayquerian by an increase in the range of RGA as well as body size (Table 6.5). In the Chapadmalalan, Ensenadan, and Lujanian, the ecological disparity of the associations was broader, but during the Platan, a decrease occurred, caused by the extinction of the largest carnivorans at the end of the Lujanian. The standardization of the area of the morphospace by the number of taxa provided a similar result, but the Platan possessed a lower disparity than the Chapadmalalan, and the Huayquerian lower than the Deseadan (Table 6.5). The use of covered morphospace for each Age as a disparity measurement gives a similar pattern.

In summary, the analyzed faunas, whether consisting of sparassodonts, carnivorans, or in combination, revealed a well-structured guild organization within the morphospace defined by diet and body mass, suggesting low intraguild

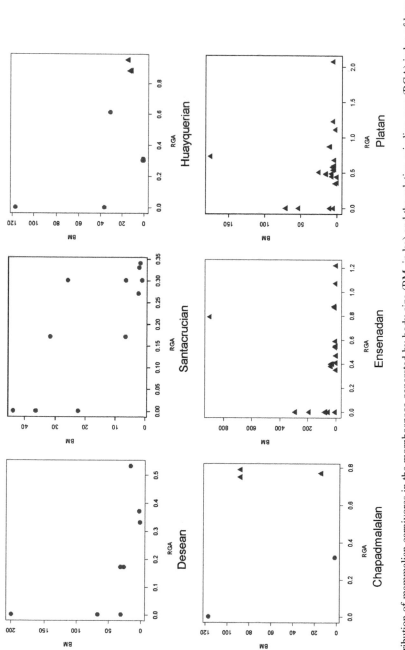

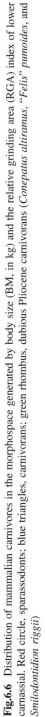

Fig. 6.6 Distribution of mammalian carnivores in the morphospace generated by body size (BM, in kg) and the relative grinding area (RGA) index of lower carnassial. Red circle, sparassodonts; blue triangles, carnivorans; green rhombus, dubious Pliocene carnivorans (*Conepatus altiramus*, "*Felis*" *pumoides*, and *Smilodontidion riggii*)

competition (see also Prevosti and Vizcaíno 2006; Prevosti and Martin 2013; Ercoli et al. 2014). These faunas become more diverse from the late Miocene at least, due to the appearance of small and large omnivores as well as hypercarnivores (including saber-toothed sparassodonts), but especially in the Pleistocene (saber-toothed cats, giant bears, large hypercarnivorous canids, as well as other living ecological carnivoran types) (see Chap. 4).

The inclusion of *"Felis" pumoides*, *Smilodontidion riggii*, and *Conepatus altiramus* in these Chapadmalalan does not modify the results, which show a good ecological separation among predators (Fig. 6.6; Table 6.5).

6.4   Carnivorans Versus Sparassodonts: Competition?

Several authors have suggested that carnivorans competed with sparassodonts, and that the former's immigration to South America during the GABI caused the latter's extinction (e.g., Simpson 1950, 1969, 1971, 1980; Patterson and Pascual 1972; Savage 1977; Werdelin 1987; Wang et al. 2008). This interpretation was informed by the ecological equivalence thought to exist between these clades using general descriptions of their anatomy (e.g., Marshall 1977, 1978, 1979, 1981), but does not consider direct quantitative analysis nor possible overlap in time of key clades in both groups. However, competition hypothesis was questioned by different authors (Marshall 1977, 1978; Reig 1981; Bond 1986; Pascual and Bond 1986; Goin 1989, 1995; Ortiz Jaureguizar 1989, 2001; Marshall and Cifelli 1990; Alberdi et al. 1995; Forasiepi et al. 2007; Forasiepi 2009; Prevosti et al. 2013; Zimicz 2014).

Diet, locomotion, and body mass have been the major paleoecological features analyzed for sparassodonts, with inferences primarily based on qualitative comparisons between marsupials and carnivorans. Based on dental features, hathliacynids and some borhyaenoids (*Prothylacynus patagonicus* and *Lycopsis torresi*) were considered to have been predominantly omnivorous (Marshall 1977, 1978, 1979, 1981). Hathliacynids were compared with didelphids, mustelids, or canids, while *Prothylacynus patagonicus* and *Lycopsis torresi* were compared with ursids and procyonids. Large borhyaenids (*Borhyaena tuberata*, *Acrocyon sectorius*, and *Arctodictis munizi*) were compared with canids and felids, being possibly able to break bones, and thus resembling living scavengers (Marshall 1977, 1978; Argot 2004a; Forasiepi et al. 2004). Thylacosmilidae, at the time represented only by *Thylacosmilus atrox,* was compared with saber-toothed cats (Machairodontinae; Patterson and Pascual 1972; Marshall 1976, 1977, 1978), due to the hypertrophy of the upper canines and associated cranial features. More recently, based on the shape of the skull and jaw, *T. atrox* was compared with nimravids (*Barbourofelis*; Prevosti et al. 2010).

More recently, statistical inferences are consistent with idea that about 90% of sparassodonts were narrowly hypercarnivorous (Wroe et al. 2004; Zimicz 2012, 2014; Prevosti et al. 2013; López-Aguirre et al. 2017; Table 3.2). Even the molar structure of hathliacynids and some borhyaenoids with small talonids (e.g., *Pseudothylacynus*, *Lycopsis*, *Prothylacynus*, and *Pseudolycopsis*) are within the range of living hypercarnivores. This feature is more evident in proborhyaenids (e.g., *Arminiheringia*, *Callistoe*, *Proborhyaena*), borhyaenids (e.g., *Australohyaena*, *Arctodictis*, *Borhyaena*), and *Thylacosmilus*, all of which virtually lack talonids and thus resemble hypercarnivorous Felidae and Nimravidae (Table 3.2). The hypercarnivorous association of sparassodonts contrasts with modern and past carnivoran communities, which are constituted by a larger proportion of omnivores and mesocarnivores (e.g., Van Valkenburgh 1999, 2007).

Reduced disparity in dental morphology within Sparassodonta in comparison to carnivorans was explained by the presence of phylogenetic constraints associated with the pattern of tooth replacement in metatherians (Werdelin 1987; see Goswami et al. 2011 for a different view). During ontogeny, lower molars erupt successively and occupy the mechanically optimal position at the middle of the jaw, until the mandible reaches adult size and the m4 assumes the most favorable site. Consequently, during development, each lower molar functions as a carnassial, at least temporarily. This form of carnassial specialization overtook specializations for other activities, such as grinding and crushing food in a typical mortar-and-pestle arrangement, as in the case of the m2–m3 in Carnivora (Werdelin 1987). However, the dentition of some sparassodonts exemplifies different evolutionary strategies to circumvent this plausible phylogenetic constraint. For example, it has been suggested that the last upper premolar of thylacosmilids (i.e., *Thylacosmilus* and *Patagosmilus*) is in fact the deciduous premolar at this locus, retained in the adult dentition (Goin and Pascual 1987; Forasiepi and Carlini 2010). This unusual ontogenetic condition, together with the hypselodont–hypertrophied upper canine, is an example of heterochronic shifts that circumvent possible constraints, increasing the morphological disparity of taxa via developmental mechanisms (Forasiepi and Sánchez-Villagra 2014).

Broad talonids represent the plesiomorphic condition for Sparassodonta. Consequently, current phylogenies tend to recover sparassodonts with broad talonids (e.g., *Patene simpsoni*, *Hondadelphys fieldsi*, *Stylocynus paranensis*) in a basal position in the tree (Forasiepi 2009; Engelman and Croft 2014; Forasiepi et al. 2015; Suarez et al. 2015). The late Miocene (Huayquerian) *Stylocynus paranensis* has been classically compared with omnivorous taxa, such as ursids and procyonids (Marshall 1978); however, cusps and crests are sharper in this species than in those carnivorans (e.g., Marshall 1979; Babot and Ortiz 2008). As a result, RGA values are lower in *Stylocynus* than in *Cyonasua* spp. which were contemporaneous in age (Tables 6.1 and 6.2; Fig. 6.6).

The postcranial anatomy of sparassodonts (Argot 2003a, b, 2004a, b, c; Prevosti et al. 2012; Ercoli et al. 2012, 2014) was quite generalized compared to carnivorans, allowing for unspecialized forms of arboreal (e.g., *Pseudonotictis*), scansorial (e.g., *Cladosictis*, *Prothylacynus*), and terrestrial activity (e.g., *Thylacosmilus*, *Borhyaena*), but without marked adaptations for cursoriality, as in canids or felids. The combination of a specialized hypercarnivorous dentition and a generalized postcranium is not commonly found among carnivorans, with the exception of some mustelids (e.g., wolverine) and some viverrids (e.g., civet).

Our revised dataset accords with the analysis of Prevosti et al. (2013), especially with regard to the temporal overlap between Sparassodonta and Carnivora during the Huayquerian–Chapadmalalan (Fig. 6.6). Our new body size estimates for *Cyonasua* (12–15 kg) place it among medium-sized carnivorans rather than small ones, as classified by Prevosti et al. (2013), but this change does not increase the ecological overlap with sparassodonts (Fig. 6.6). *Cyonasua* spp. remain as omnivores in our classification, which means that their only potential Huayquerian competitor would have been *Stylocynus* (with a body mass of 31 kg; Chaps. 3 and 4). However, note that based on actualistic evidence (Dickman 1986; Palomares and Caro 1999; Donadio and Buskirk 2006; Oliveira and Pereira 2014), the larger sparassodont would have been dominant in competitive situations with the smaller carnivoran. A similar conclusion has been recently suggested by López-Aguirre et al. (2017) based on a slightly different database averaged to generic level, and multiple regression and beta diversity analyses.

Another possibility is that competitive displacement between these carnivore clades might be masked by the imperfections of the fossil record, which do not allow us to ecologically differentiate sparassodonts and carnivorans with inferred similar habits (Chap. 5). We cannot completely dismiss the possibility that future records will reveal that sparassodonts become extinct later in time than the mid-Pliocene, or, alternatively, that large hypercarnivorous or omnivorous carnivorans appeared earlier in South America. Here we test the following propositions: (1) The sampling effort has affected recognition of apparent ecological overlap during the late Miocene–Pliocene (see Chaps. 5 and 2) the inclusion of doubtfully Pliocene taxa (*"Felis" pumoides*, *Smilodontidion riggii*, and *Conepatus altiramus*) affects interpretation of first arrivals of these carnivoran ecological types.

In Chapter 5, we recognized important biases in the SA fossil record. In order to test the possibility that the biases are obscuring the diversity and ecological patterning between sparassodonts and carnivorans, we compared differences in the abundance of observed ecological classes (a combination of diet and body size) in each clade during the Huayquerian–Chapadmalalan (i.e., large hypercarnivores, small hypercarnivores, medium omnivores, large omnivores), using a randomly generated ratio to obtain the probability that an observed instance is larger than the random estimate. This was performed twice, once using the proportion of the number of localities in each age, and once using the number of specimens, up to the size of each resampling. The observed abundance difference between sparassodonts and carnivorans in the large hypercarnivore category is highly significant ($p = 0.0001$) for the Huayquerian and Chapadmalalan, but not for the Montehermosan ($p > 0.05$). This means that the randomly expected difference is lower than the observed one. When the Santacrucian fauna is used for the resampling, the observed difference in the number of specimens of large hypercarnivores is significant for the Huayquerian ($p < 0.0135$) and Chapadmalalan ($p < 0.005107$). The observed difference for medium omnivores was always lower than randomly expected ($p < 0.01$). Similar results were obtained for small

hypercarnivores, except for the Chapadmalalan when Huayquerian or Friasian–
Chasicoan faunas were used in the resampling (with both sample proxies), and for
the Huayquerian when Friasian–Chasicoan faunas were used in the resampling
(using the number of localities), with a non-significant pattern ($p > 0.05$). We use
the same procedure to test how the limited fossil record of the Marplatan Subages
could mask the overlap between large and small hypercarnivores (the ecological
types represented by the last sparassodonts) due to potential persistence of
sparassodonts or to an earlier immigration of large hypercarnivorous carnivorans
during the Marplatan. The Montehermosan includes very few localities; consequently, the random expectation regarding the abundance of large hypercarnivores
among sparassodonts versus carnivorans is mostly not significantly different from
the observed absence of them in the Marplatan Subages. The only exceptions are
the Vorohuean and Sanandresian when the Friasian–Colloncuran faunas are used in
the resampling; in these cases, values are significantly lower ($p = 0.03$). On the
other hand, the random difference between small hypercarnivores is significantly
greater ($p = 0.02$) than the observed one.

With regard to the inclusion of dubious carnivorans in the Chapadmalalan (two
large hypercarnivores and one small omnivore), the difference between small
omnivores is not significant, except when the Friasian–Chasicoan faunas are used in
the resampling. In this case, the difference is significantly smaller than the random
expectation ($p < 0.005$). By contrast, the large hypercarnivore difference is highly
significantly larger than the random expectation ($p = 0.0001$), except when the
Santacrucian and number of localities are used. However, the exclusion of only one
of the dubious large hypercarnivores results in a significantly larger value
($p = 0.0344$). This is relevant because even if a Pliocene age is accepted for "*Felis*"
pumoides, it is not possible to allocate it in any specific age within the Pliocene,
Chapadmalalan, or otherwise.

These results, in combination with arguments and analyses presented above and
in Chapter 5, suggest that competition between carnivorans and sparassodonts in
the Huayquerian–Chapadmalalan was absent, because the observed difference of
hypercarnivores is significantly larger than random expectation. A similar conclusion can be made for large omnivores in the Huayquerian. One caveat is that faunal
structure during the late Miocene–Pliocene is different from that of the Pleistocene
and older Neogene epochs. The results obtained for some ecological classes (e.g.,
between-clade differences significantly larger than random for medium omnivores
and/or small hypercarnivores) are congruent with this possibility (Prevosti and
Soibelzon 2012; Prevosti et al. 2013). Although it is difficult to unambiguously
predict the effect of this caveat, if the inferred late Miocene–Pliocene absence of
hypercarnivore carnivorans and limited sparassodont diversity accurately mirrors
past occupancy of the predator guild, our sense of an absence of competition is only
strengthened.

Another relevant result of these analyses is that Ages with very small sample
sizes (either in number of specimens or localities; e.g., Montehermosan and
Marplatan Subages) provide only ambiguous or non-significant results and thus do
not permit rejection of between-clade competition. This is especially true for the

Marplatan, where low sampling may mask an overlap of sparassodont and carnivoran hypercarnivores. Nonetheless, as discussed in Chap. 5, the available evidence is insufficient to support the coexistence of these morphotypes and is in fact more congruent with absence of competition.

If the presence of the three dubious carnivoran taxa record in the Pliocene is eventually confirmed, judging from actualistic ecological studies of intraguild competition (Dickman 1986; Palomares and Caro 1999; Donadio and Buskirk 2006; Oliveira and Pereira 2014), *Smilodontidion riggii* (=*Smilodon* sp.) is the only taxon that could have actively displaced *Thylacosmilus*. However, these sabertooths did not share the same morphotype or body mass (Wroe et al. 2013), and this difference might have influenced their prey selection and predatory habits (Chap. 3). In turn, "*Felis*" *pumoides* was comparable in size to a small puma (*Puma concolor*) and thus much smaller than *Thylacosmilus*.

Summing up, data and analyses are more congruent with the absence of competition between sparassodonts and carnivorans. More fieldworks and additional analyses using new approaches to test competition, clade diversification, and sampling effort (Liow and Finnarelli 2014; Pires et al. 2015; Silvestro et al. 2015; Finarelli and Liow 2016) potentially will provide new insights to test this interpretation.

6.5 Terror Birds and Carnivorous Didelphimorphian: Other Cases of Competition?

Alternative competitive exclusion hypotheses affecting sparassodonts include the possible role of "terror birds" (Phorusrhacidae; Marshall 1977, 1978, *partim*; Marshall and Cifelli 1990; see also Croft 2006) and didelphimorphian marsupials with carnivorous dentitions (Marshall 1977, 1978, *partim*; Goin 1989; Goin and Pardiñas 1996).

Terror birds were flightless, cursorial predators that were present in South America during most of the Cenozoic and included small to gigantic forms (Degrange et al. 2012). The group also reached North America at the end of the Neogene and was represented in the Pliocene by *Titanis walleri* (MacFadden et al. 2007).

Appraisal of the diversity of Phorusrhacidae in South America (Alvarenga and Höfling 2003; Degrange et al. 2012) indicates that the number of species was low during the Cenozoic, with peaks of only four species during Deseadan, Santacrucian, and Huayquerian time (Fig. 6.1). The latest records with reliable stratigraphic data are from the Chapadmalalan (Degrange et al. 2015), but material from putative Pleistocene beds have been found in Uruguay (Tambussi et al. 1999; Agnolin 2009). The most recent of these seemingly include a phorusrhacid from the Late Pleistocene (Lujanian) (Alvarenga et al. 2009; Jones et al. 2016).

Marshall and Cifelli (1990; see also Marshall 1977, 1978) suggested that terror birds may have displaced large sparassodonts during the Paleogene, resulting in the disappearance of proborhyaenids. Yet later, during the late Miocene–early Pliocene when savannas, pampas, and open environments expanded, advantages for cursorial carnivores like terror birds should have been even greater. However, biochron data for phorusrhacids and diversity peaks are roughly similar to those of sparassodonts (Fig. 6.1), indicating that competition was unlikely between these clades and, minimally, that niches were at least partially segregated (see also Marshall 1978; Bond and Pascual 1983; Argot 2004b; Prevosti et al. 2013; Ercoli et al. 2014).

Birds represent the second major group of terrestrial predators in the early Miocene Santacrucian vertebrate association, with four phorusrhacids, one cariamid, one anatid, and two falconids (Degrange 2012; Degrange et al. 2012). The largest species was the probable scavenger *Brontornis burmeisteri*, an anseriform estimated to have weighed more than 300 kg (Tonni 1977). Among "terror birds," the largest phorusrhacid was *Phorusrhacos longissimus*, about 100 kg and an active predator on large prey (Fig. 6.2). It possessed a rigid cranium, high bite force, and stocky neck and limbs (Degrange 2012; Degrange et al. 2012). Smaller phorusrhacids were represented by *Patagornis marshi* (30 kg), *Psilopterus lemonei* (10 kg), and *Psilopterus bachmanni* (4.5 kg). Comparing the body sizes of phorusrhacids and sparassodonts, *Patagornis marshi* is within the range of *Acrocyon sectorius* and *Prothylacynus patagonicus*. If the ecological impact of body size (translated into prey size) can be taken as a comparable proxy between birds and mammals, potential competition becomes a possibility (Ercoli et al. 2014). However, long term competition (Croft 2001, 2006) and locomotory differences would have favored the adoption of different predatory strategies on the part of these taxa (Ercoli et al. 2014).

Sparassodonts were a group that failed to develop a strict cursorial morphotype. While bone-crackers and other carnivores are commonly ecologically associated with other more efficient predators that partially consume and abandon the carcasses (Viranta 1996; Argot 2004b), this latter role could have been occupied by phorusrhacids. Consequently, it seems highly probable that these terrestrial birds occupied an ecological niche different from that of the sparassodonts (Argot 2004b; Ercoli et al. 2014; López-Aguirre et al. 2017), and that consequently the phorusrhacids did not force their extinction.

Similarly, other non-mammalian predators were in a minority in terrestrial ecosystems. The diversity of sebecid crocodiles and madtsoiid snakes was low during the Cenozoic (Gasparini 1996; Albino 1996; Paolillo and Linares 2007; Riff et al. 2010), and their extinction preceded that of sparassodonts (Fig. 6.1). During the late Miocene–Pleistocene, several different carnivorous morphotypes developed within didelphimorphian marsupials, some of which are still represented in the living fauna. These were distributed in two main groups: Sparassocynidae (*Hesperocynus dolgopolae*, *Sparassocynus bahiai*, *S. derivatus*, *S. heterotopicus*) and Didelphidae (*Didelphis albiventris*, *D. crucialis*, *D. reigi*, *D. solimoensis*, *Hyperdidelphys dimartinoi*, *H. inexpectata*, *D. parvula*, *D. pattersoni*, *Lestodelphys halli*, *Lutreolina crassicaudata, L. tracheia, Thylatheridium cristatum, T. chapadmalensis, T. hudsoni, T. pascuali, Thylophorops chapadmalensis, Th. lorenzinii,* and *Th. perplanus*) (Zimicz 2014). Sparassocynidae is an extinct family of carnivorous didelphimorphians (<1 kg in body mass; Zimicz 2014). Didelphidae includes omnivorous to carnivorous forms. The latter filled the size range from small (body sizes of a few grams) to medium (sizes up to about 10 kg; Zimicz 2014). The last hathliacynids (*Borhyaenidium altiplanicus, B. musteloides, B. riggsi, Notocynus hermosicus, Notictis ortizi*) overlapped with some carnivorous didelphimorphians during the Huayquerian–Chapadmalalan (Sparassocynidae, *Didelphis, Hyperdidelphys, Lutreolina, Thylateridium, Thylophorops*). This temporal and body size overlap, together with the terrestrial to scansorial habits of several extinct carnivorous didelphids, suggested a possible competitive interaction between small- and medium-sized hathliacynids and the carnivorous didelphimorphians (Marshall 1977, 1978, *partim*; Goin 1989; Goin and Pardiñas 1996; Forasiepi 2009). However, mesocarnivorous and hypocarnivorous, but not hypercarnivorous, diets have been inferred for the didelphimorphians occurring within the period of temporal overlap (Zimicz 2014). These differences point to the existence of a potential niche segregation between these two metatherian groups, without competitive displacement (Prevosti et al. 2013) or passive replacement (Zimicz 2014).

Correlation analysis (Spearman correlation coefficient) of the diversity of sparassodonts versus that of phorusrhacids, sebecids, or madtsoiid snakes, individually and in combination, support these interpretations, because most correlations are not significant ($p > 0.05$). The exception is total diversity of non-mammalian predators ($R = 0.60$, $p < 0.05$). However, the correlation is positive rather than negative, as would be expected in a competitive exclusion scenario. There are only three Ages in which sparassodonts and carnivorous didelphids overlap (Huayquerian, Montehermosan, and Chapadmalalan), making correlation analysis irrelevant, but in any case the observed data do not show a clear pattern that could be seen to support competition between sparassodonts and carnivorous didelphimorphians (Prevosti et al. 2013; Zimicz 2014).

6.6 Paleoenvironments and Faunal Changes: Do These Correlate?

The relevance of environmental changes and their impact on terrestrial mammalian communities in connection with the demise of sparassodonts occasionally has been addressed (see also Marshall 1977, 1978, *partim*; Marshall and Cifelli 1990, *partim*; Forasiepi et al. 2007; Prevosti et al. 2013). As documents in Chapter 2 (see also Pascual and Odreman Rivas 1971; Patterson and Pascual 1972; Pascual et al. 1985; Pascual and Ortiz Jaureguizar 1990; Zachos et al. 2001; Barreda and Palazzesi 2007; Barreda et al. 2008; Dozo et al. 2010), Neogene global climate change affected South America. In the middle Miocene–Recent, nonlinear but gradual global temperature decrease has been the rule, correlated with the establishment of permanent ice sheets in western Antarctica and the onset of glaciation in high-latitude Northern Hemisphere (Zachos et al. 2001). Rejuvenation of Andean orogeny during the Neogene clearly modified prevailing conditions in the physiography, climate, and biota of South America. Marine transgression, lacustrine development, and widespread fluvial incision would have affected previous biogeographic patterns during the middle and late Miocene (Campbell et al. 2006; Cozzuol 2006; Latrubesse et al. 2007; Marengo 2015).

For some regions of South America (e.g., Patagonia), there is evidence of the onset of desertification, continuing up to the present. This process featured replacement of forest by more open and xerophytic vegetation in the middle–late Miocene; in particular, open dry forest, composed of *Schinus* (pepper trees), *Prosopis* (algarrobos), and *Celtis* (nettle trees) with shrubs of Ephedraceae and Asteraceae, is known for the late Miocene (Barreda and Palazzesi 2007; Barreda et al. 2008; Dozo et al. 2010; Fig. 6.8). Floras featuring mixtures of C3–C4 species have been recovered from localities situated between 21° and 35°S in Bolivia and Argentina, suggesting the existence of extensive grasslands by 8 Ma (MacFadden et al. 1996) throughout a large part of South America.

In the Neogene, the observed climatic and floral changes in Patagonia also impacted mammalian paleocommunities, which experienced important compositional changes, including the extinction of some native lineages (Pascual and Odreman Rivas 1971; Patterson and Pascual 1972; Pascual et al. 1985; Pascual and Ortiz Jaureguizar 1990). Several South American "native ungulates" (e.g., Astrapotheria, Leontinidae, Adianthidae, Notohippidae) became extinct during the middle Miocene, although Toxodontidae and certain xenarthran taxa experienced a radiation (Megalonychidae, Megatheriidae, and Mylodontidae; Marshall and Cifelli 1990). On the whole, "native ungulates" experienced a continuous decline from the middle Miocene onward, with a steep reduction in the Pliocene and final disappearance during the Pleistocene–Holocene (Marshall and Cifelli 1990; Bond et al. 1995). In the Pampean Region, where the vertebrate fossil record for the late Miocene–Quaternary is relatively good, a faunal turnover has been detected for the mid-Pliocene (e.g., Kraglievich 1952; Tonni et al. 1992). Some authors have argued that this was related to environmental changes triggered by Andean orogeny rather than competition with North American immigrants (e.g., Ortiz Jaureguizar et al. 1995; Cione and Tonni 2001), but a more recent hypothesis suggests that a meteor impact in the Pampean Region during the late Chapadmalalan (ca. 3.3 Ma) drove this faunistic change (Schultz et al. 1998; Vizcaíno et al. 2004).

In this context, the decrease in sparassodontan diversity and their extinction in the mid-Pliocene appear to be part of a more general phenomenon of widespread faunal turnover independent of clade affiliation, life history, or major adaptations (Prevosti et al. 2013; Fig. 6.7). Sparassodonts, especially larger ones with a dietary specialization toward hypercarnivory, may have been more vulnerable to extinction than non-hypercarnivorous species (cf. Van Valkenburgh et al. 2004). Additionally, if sparassodonts had imperfect homeothermy and lactation limited to the rainy season, as do living marsupials (McNab 1986, 2005, 2008; Green 1997;

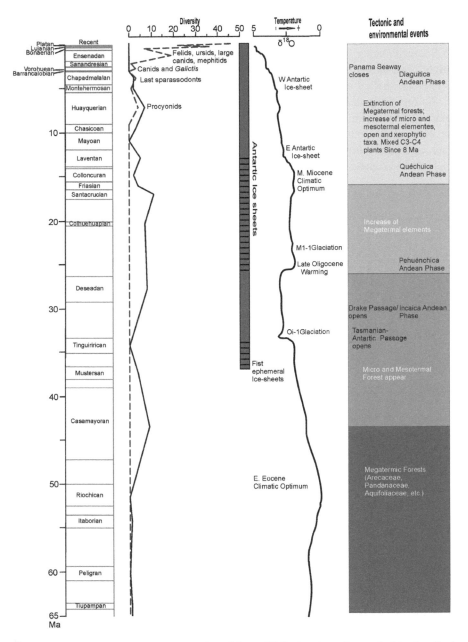

Fig. 6.7 South American Ages, sparassodont (blue solid line) and carnivoran (red broken line) diversity, temperature curve, climatic, environmental, and tectonic events (Modified from Prevosti et al. 2013). Colored boxes correspond to different floral changes (darker is older). Antarctic ice sheets: Dashed bar represents minimal ice (<than 50% of present ice volume), while gray represents full glaciation (>50% of present ice volume; Zachos et al. 2001)

Krockenberger 2006), the decrease in temperature and increase in aridity during the late Miocene when xerophytic vegetation and open environments were established in South America (Barreda and Palazzesi 2007; Barreda et al. 2008; see Chap. 2) might have influenced the decline of the group. This is a plausible argument, but it is unclear whether extrapolation from living marsupials to sparassodonts is justified.

To test the potential role of climatic and biological factors in the evolution of sparassodonts, we performed simple correlation tests (Spearman coefficient) between global temperature (median, minimum, maximum, and range of $\delta^{18}O$ for each age, taken from Zachos et al. 2001; Table 6.6) against a range of variables including systematic diversity, estimated body mass, body size abundance, dietary

Table 6.6 Global temperatures and prey body mass for South American taxa

Age	T med	T min	T max	T range	Prey med	Prey min	Prey max	Prey range
Tiupampan	0.62	0.45	0.69	0.24				
Peligran	0.63	0.38	0.86	0.48				
Itaboraian	0.22	−0.38	0.45	0.83				
Riochican	−0.12	−0.28	0.14	0.42	50	48.68	400	351.32
Casamayoran	0.96	0	1.48	1.48				
Mustersan	0.735	0.35	1.21	0.86				
Tinguirirican	1.82	1.6	3	1.4				
Deseadan	2.69	2.33	2.99	0.66	204	45	3500	3455
Colhuehuapian	2.08	1.96	2.24	0.28				
Santacrucian	1.69	1.4	2.06	0.66	121.26	70	1021.63	951.63
Friasian	1.79	1.51	2.01	0.5				
Colloncuran	1.76	1.43	2.21	0.78				
Laventan	2.34	1.98	2.81	0.83				
Mayoan	2.58	2.33	2.86	0.53				
Chasicoan	2.74	2.43	3.06	0.63				
Huayquerian	2.93	2.43	3.5	1.07	305.5	23.24	3000	2976.76
Montehermosan	3	2.59	3.46	0.87	300	27	3397	3370
Chapadmalalan	3.1	2.71	3.62	0.91	400	50	3000	2950
Barrancalobian	3.28	2.96	3.66	0.7	250	27	1344	1317
Vorohuean	3.41	2.97	3.88	0.91	160	27	3950	3923
Sanandresian	3.64	3.1	4.16	1.06	731	50	7580	7530
Ensenadan	3.89	3.07	4.98	1.91	600	50	7580	7530
Bonaerian	4.18	3.24	4.84	1.6				
Lujanian	4.07	3	4.91	1.91	600	50	7580	7530
Platan	3.325	3.23	3.39	0.16	35	16	300	284

Temperatures were taken from the Zachos et al. (2001); $\delta^{18}O$ curve and prey body mass from Vizcaíno et al. (2004, 2012), Villafañe et al. (2006), Reguero et al. (2010), Cassini et al. (2012), and Forasiepi et al. (2016) (raw data available on request). *T* temperature; *Prey* prey body mass (kg); *med* median; *min* minimum; *max* maximum; *range* observed range

classes, and median, minimum, and maximum RGA of lower carnassials (Chaps. 3 and 4; Tables 6.1, 6.2, 6.3 and 6.4). For a limited set of faunas, we also included the median, minimum, and maximum body size of medium and large prey species (Table 6.6). Lazarus taxa are alternately included or excluded from the analysis of systematic diversity, body size abundance, climate, and dietary classes.

By including Lazarus taxa, we try to ameliorate bias in the fossil record. In contrast, the inclusion of ages lacking sparassodont fossils or Lazarus taxa explores the possibility that these absences are real—a hard assumption to accept—given that ghost lineages indicate that the taxa must have existed, even if their fossils are (so far) unknown. Due to the large number of correlations, we adjust p values using the False Discovery Rate procedure (Benjamin and Hochberg 1995).

Total diversity of sparassodonts is positively correlated with temperature (Spearman R between 0.40 and 0.53, $p < 0.05$). However, it is only when Lazarus taxa are included that this correlation is significant at the p level indicated by the FDR correction ($p = 0.0075$), when minimal temperature shows a correlation of 0.53. Also, when Lazarus taxa are included, the abundance of omnivores and small taxa shows a positive correlation with temperature (Spearman R 0.63–0.67 and 0.54–0.59, respectively, $p < 0.0075$), but when ages lacking omnivores and/or small taxa are excluded, these correlations become non-significant even at $p = 0.05$. The same pattern is observed when Lazarus taxa are omitted, but in this case the correlation is lower (Spearman R 0.41–0.48, only significant at $p = 0.05$). Body mass and RGA reveal a contrasting pattern, especially when Lazarus taxa are included, with the first variable being negatively correlated (Spearman R 0.50–0.82) and the second positively correlated (Spearman R 0.50–0.74) with temperature. Most of these correlations are significant at the p level established by the FDR correction ($p = 0.0075$). The same pattern was found when Lazarus taxa are excluded, but few correlations are significant after FDR correction (e.g., median body mass against median and minimum temperatures).

These correlations imply a possible connection between temperature change and sparassodont evolution and extinction during the Cenozoic. Diversity, RGA, and abundance of omnivores and small taxa all increase with rising global temperature, while body size (median and maximum) increases with declining temperature. The similar response of RGA and omnivore abundance is consistent with this picture, since increase in the first variable indicates that sparassodont faunas are less carnivorous (or, alternatively, more omnivorous).

In relation to prey body mass distribution, the only significant correlations are between range and maximum prey body size, as well as and range and maximum sparassodont body mass, with or without Lazarus taxa or using the FDR correction. Maximum sparassodont body size has the largest correlation (Spearman R 0.94–0.95, $p < 0.006$). Range of sparassodont body mass also is positively correlated (Spearman R 0.82), but is only marginally significant ($p = 0.049$).

Our results are congruent with the analyses of Zimicz (2012, see also Goin et al. 2016). They also detected a positive relationship between diversity and omnivore abundance and temperature, but no correlation between body mass and temperature (Zimicz 2012). In our results, the increase of body size with temperature decrease could be seen as another example of the Bergmann and/or Cope "rules," but this hypothesis should be tested in a phylogenetic context (e.g., Gould and MacFadden 2004). On the other hand, maximum prey size apparently influenced maximum sparassodont body size, since a highly positive relationship was detected, at least in regard to the faunas included in this analysis. Maximum prey body size apparently decreased in the Barrancalobian (from 3 tons in the Chapadmalalan to ca. 1.5 tons), though this could be partly due to a bias present in this Age. Later prey body size experienced an increase to ca. 4 tons in the Vorohuean and to ca. 7.5 tons in the Sanandresian–Lujanian, when the largest body size is achieved (Table 6.6). Thus, if sparassodonts did not become extinct in the Chapadmalalan, we would expect an increase in their maximum body size after the Vorohuean. But several factors could modify this expectation, including a threshold in body size above which prey are invulnerable to predation and, consequently, irrelevant to sparassodont evolution. The large increment in prey body mass observed since the Sanandresian is generated by the immigration of gomphotheres, a lineage, and morphotype that did not evolve together with sparassodonts. On the other hand, several native lineages ("native ungulates", glyptodonts, ground sloths) increase their body mass in the Pleistocene, the largest being the ground sloth *Megatherium americanum* that had a body mass of ca. 4 tons (Vizcaíno et al. 2012), 1 ton larger than the maximum recorded in the Chapadmalalan. Thus, even excluding gomphotheres, at least a modest increase in maximum sparassodont body size should be expected if other factors were not involved.

Inclusion of Lazarus taxa significantly affected correlations. This may be taken as an indication that the use of methods correcting for bias in the fossil record might recover more relationships between sparassodont diversity and variables concerned with environmental and faunistic factors. Reduction of global temperatures during the late Miocene–Pleistocene may have had a negative impact on sparassodonts, probably in correlation with the decline and extinction of South American native ungulate lineages. Obviously, these results, based as they are on elementary correlation analyses, should be taken as preliminary working hypotheses that should be tested with more robust methods and a wider faunistic sample, including ones that control for bias (e.g., Liow and Finarelli 2014; Silvestro et al. 2015; Pires et al. 2015; Finarelli and Liow 2016).

6.7 The Extinction of Sparassodonta: Competition with Incoming Carnivorans, Environmental and Faunal Changes, or a Combination Thereof?

Our data and analyses support the hypothesis that there was no competition between sparassodonts and carnivorans inhabiting South America in the late Miocene–Pliocene, because they did not overlap in ecological space. Dubious records of some carnivorans in the Pliocene do not meaningfully support ecological competition between these clades. Other evidence points to environmental as well as more general change in the South American mammalian fauna forced the decline and demise of sparassodonts in the Pliocene. The available data are robust for the decline of sparassodonts since the Huayquerian, but for the final loss of this clade, the information is more ambiguous since Marplatan is not a well-sampled Age and/ or has a poorer fossil record.

A similar conclusion has been recently supported by López-Aguirre et al. (2017), using a multivariate statistical approach. In their view, the extinction of the Sparassodonta is related to faunal changes and ecological interaction with other non-predator mammals. The authors provides less support to environmental changes (climate and Andean orogeny); however, the proxies used in their study to record Andean rising only considered one section of the Andes, and they use a global climatic scale that does not track changes at local scale.

We think that the hypothesis that Sparassodonta became extinct due to environmental and faunal changes has the epistemological advantage of being more vulnerable to falsification than its alternatives, such as interclade competition. We hope that additional targeted fieldwork and the application of more refined methodologies will provide useful tests (e.g., Liow and Finarelli 2014; Silvestro et al. 2015; Pires et al. 2015; Finarelli and Liow 2016).

6.8   Carnivoran Impact on Pleistocene Faunas: Giant Makers?

The immigration of placental predators to SA during the GABI impacted on the communities of native prey, in some cases triggering their extinction (e.g., Marshall 1988). If correct, this illustrates on trophic interactions intervening and modulating the dynamics of macroevolutionary processes. In this sense, Faurby and Svenning (2016) have recently demonstrated that the asymmetry in the exchange of mammals during the GABI, with NA clades more successful in SA, could be explained by differential susceptibility to predation pressure between animals from the two continents. But, does another biological response exist against the arrival of several carnivoran lineages into SA during the early Pleistocene? In this regard, Soibelzon and colleagues (Soibelzon et al. 2009, 2012; Zurita et al. 2010) argued that the presence of large carnivorans produced an adaptive response in several native herbivores due to predation pressure created by the newcomers. Gigantism observed in several lineages during the Pleistocene (e.g., glyptodonts, ground sloths, toxodontids, and macrauchenids), all of whom had body sizes larger than 1 ton (Vizcaíno et al. 2012; Fariña et al. 2013), behavioral changes (e.g., subterranean dwelling; Soibelzon et al. 2009, 2012), and anatomical changes (e.g., thicker carapace in some large glyptodonts; Zurita et al. 2010) could be correlated with defenses against wholly new types of predators. These hypotheses were proposed to explain the coincident occurrence in the SA fossil record of the first large carnivorans and megaherbivores in the Ensenadan.

Another scenario explains the existence of a continent with a diverse community of large mammals and megamammals with few large predators. This could have given ecological space to the new Pleistocene carnivoran immigrants. This scenario is congruent with the interpretation that the largest Pleistocene carnivorans (e.g., *Arctotherium* spp., *Smilodon populator*, *Theriodictis platensis*, *Protocyon* spp., and *"Canis" gezi*) apparently had their origins in South America (Figueirido and Soibelzon 2009; Soibelzon and Schubert 2011; Prevosti and Soibelzon 2012; Mitchell et al. 2016).

These two hypotheses are not mutually exclusive. South America hosted large mammals and megamammals during the Pleistocene, prior to the arrival of the large carnivorans (Vizcaíno et al. 2012), which could have provided ecological space for the development of large-bodied predators. In turn, the pressure exerted by these predators could have favored larger body sizes among herbivores, in an arm-race-like scenario (cf. Van Valen 1973; Dawkins and Krebs 1979). Alternatively, this faunistic change and increasing body mass could be governed by climatical–environmental changes (cf. Barnosky 2001; Raia et al. 2005).

Utilization of caves or dens in SA terrestrial environments by autochthonous mammals predated the arrival of large carnivorans (De Elorriaga and Visconti 2002; Genise et al. 2013; Cardonatto et al. 2016), but substantial enlargement of such dwellings postarrival clearly reflects an increase in body sizes. Large caves of up to 1.5 m width, attributed to ground sloths, armadillos, glyptodonts, or mesotheriids, are frequently encountered in late Miocene rocks (e.g., Cerro Azul Formation: De Elorriaga and Visconti 2002; Cardonatto et al. 2016). Quaternary paleocaves surpassed this value, attaining in some cases as much as ca. 2 m in width (Vizcaíno et al. 2001; Dondas et al. 2009; see also Fariña et al. 2013).

In 1996, Fariña and collaborators proposed the hypothesis that the Late Pleistocene South American faunas were not in balance, with a larger biomass of herbivores than predators (Fariña 1996; Fariña et al. 2014; see also Fariña et al. 2013; Fig. 6.8). This conclusion was reached using allometric relationships among living mammals in relation to their estimated body masses. However, the methodology was criticized, because it did not include comparative methods to constrain phylogenetic effects in calculating allometric equations, adjustments for taphonomic biases produced by time averaging of associations, or variability observed in population density or biomass in living communities (Prevosti and Vizcaíno 2006; see also Prevosti and Pereira 2014). Based on these criticisms, another scenario was proposed: If the Late Pleistocene supported a large biomass of herbivores, then it could have also supported a large density of predators (Prevosti and Vizcaíno 2006; Fig. 6.8). In fact, in a more recent analysis, Segura et al. (2016) suggested that the structure of the Late Pleistocene faunas of Buenos Aires (Argentina) and Uruguay resembles modern faunas.

Prevosti and Vizcaíno (2006) also investigated prey–predator relationships by means of allometric equations (and, more recently, by stable isotope analyses; Prevosti and Martin 2013; Bocherens et al. 2016) and discussed intraguild

(a)   **(b)**

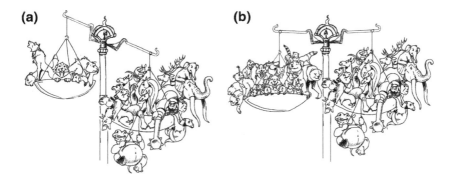

Fig. 6.8 Schematic representation of the predator–prey "imbalance" hypothesis of Fariña (1996) (**a**) and the alternative hypothesis proposed by Prevosti and Vizcaíno (2006) under which the high diversity of large mammals and megamammals supported a great abundance of predators (**b**). From Prevosti and Vizcaíno (2006)

relationships. The largest hypercarnivore occurring in SA Late Pleistocene associations was *Smilodon populator*. This taxon is the most common carnivoran found in fossil sites of this age, and its presence probably generated a top-down negative cascade affecting other predators (Palomares and Caro 1999; Donadio and Buskirk 2006; Oliveira and Pereira 2014), with the possible exception of *Arctotherium angustidens*, whose body mass could exceeded 1 ton (Chap. 4).

6.9 Carnivoran Extinctions During the Pliocene–Holocene: Intraguild Competition, Human Impact, or Another Cause Entirely?

An important change in carnivoran faunas occurred between the late Pliocene (Vorohuean) and the early Pleistocene (Ensenadan), when the procyonid "heralds" of the GABI became extinct and the carnivoran contingent of the "legions" of the GABI appeared. The extinction of the large bear-like *Chapalmalania* at the end of the Pliocene (Vorohuean) was suggested to be a consequence of the immigration and radiation of bears in the Pleistocene, ca. 1.8 Ma (Kraglievich and Olazábal 1959; Prevosti and Pereira 2014). However, the coexistence of these clades remains unconfirmed, with a gap of ca. 0.8 ka in the first appearances of these two groups in South America (Table 4.1). A recent study based on ancient DNA suggested that the origin of large scavenging bears was an independent development in North and South America (*Arctodus* and *Arctotherium*), facilitated by the absence or limited number of other large mammalian scavengers (Mitchell et al. 2016). A parallel exists in the extinction of *Cyonasua* during the early Pleistocene and the presence of new carnivorans in the Pleistocene (Soibelzon 2011), albeit with insufficient information to exclude other possibilities (e.g., environmental and faunal changes; Prevosti and Pereira 2014).

A number of extinctions occurred at the end of the Ensenadan (ca. 0.5 Ma) when 12 taxa disappeared (e.g., *Theriodictis platensis*, *Protocyon scagliarum*, *"Canis" gezi*, *Lycalopex cultridens*, *Arctotherium angustidens*, *Cyonasua* spp., *Lyncodon bosei*, *Stipanicicia pettorutii*, *Galictis hennigi*, *Smilodon gracilis*, *Homotherium venezuelensis*, *"Felis" vorohuensis*; Prevosti and Soibelzon 2012). This event affected large and small hypercarnivores and omnivores (Prevosti and Soibelzon 2012). The degree to which these losses were actually coordinated in time is not well understood; possibly, their apparently shared end-Ensenadan disappearance date may be an artifact of analytic averaging of South American Land Mammal Ages and the effect of taphonomic biases. Last appearance records for *L. bosei* and *Cy. meranii* are dated to ca. 1 Ma (Soibelzon et al. 2008), indicating that they were possibly extinct before other Ensenadan carnivorans. In addition, the Bonaerian Age has a poor fossil record which may fail to include some Ensenadan carnivorans (Chap. 5). The recent extension of the time span of *Lycalopex ensenadensis* and *Conepatus mercedensis* into the Lujanian (Chap. 4) supports this interpretation. The end-Ensenadan extinction is roughly coincident with the establishment of wider climatic cycles of 100 ka, with colder glacials and warmer interglacials since 0.9 Ma (Rutter et al. 2012) and could be related to the faunistic changes observed between this Age and the Bonaerian–Lujanian.

The Late Pleistocene–Holocene transition incorporates another mammalian extinction event that also included large carnivorans, spawning a diverse set of hypotheses over the last decades. These include environmental changes, human hunting, introduced pathogens, meteoric impact, and the combination of all or any of the above (Martin and Wright 1967; MacPhee and Marx 1997; Cione et al. 2003, 2009; Borrero 2009; Lima Ribeiro and Diniz Filho 2013; Prado et al. 2015; Surovell et al. 2015; Villavicencio et al. 2016; Metcalf et al. 2016; Bartlett et al. 2016). In South America, this extinction occurred between 12 and 7 ka BP, with the Pampean Region showing the latest completion date (Cione et al. 2003, 2009; Steele and Politis 2009; Prado et al. 2015; Villavicencio et al. 2016; Metcalf et al. 2016; Politis et al. 2016). South American carnivorans followed a similar pattern to other mammals, in that large mammals and megamammals became extinct, and most large carnivorans disappeared (i.e., saber-toothed cats, large hypercarnivorous canids, bears; Table 4.1; Fig. 6.9), leaving the puma *Puma concolor*, jaguar *Panthera onca*, and spectacled bear *Tremarctos ornatus* as the only living large carnivores on the continent (Cione et al. 2003, 2009; Prevosti and Soibelzon 2012). Small carnivorans that became extinct in the Late Pleistocene (*Lycalopex ensenadensis* and *Conepatus mercedensis*) could have disappeared before the end of the epoch (Chap. 4).

The disappearance of the large mammals and megamammals would have impacted the larger carnivorans, resulting in their mutual extinction (Prevosti and Vizcaíno 2006; Prevosti 2006; Prevosti and Martin 2013; cf. Van Valkenburgh et al. 2015). Isotopic analysis of material of *Smilodon populator* and *Protocyon troglodytes* from the Late Pleistocene of the Pampean Region indicated that they

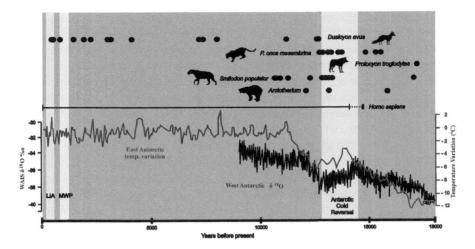

Fig. 6.9 Taxon dates (calibrated ^{14}C dates BP) of terrestrial extinct South American carnivorans (black dots), presence of humans in South America, and paleoclimate proxy (West Antarctic δ^{18}O). *LIA* little ice age; *MWP* medieval warm period; *WAIS* West Antarctic ice sheet. East Antarctic refers to Vostok core, and temperature variation to past minus present temperature (°C). Modified from Metcalf et al. (2016)

hunted in more open environments than the jaguar and could have suffered from the warmer and wetter climate of the Holocene (Bocherens et al. 2016). Recently, Villavicencio et al. (2016) considered that large carnivorans became extinct before large megamammals in southern Patagonia, due to the appearance of humans. However, when doubling the number of radiometric dates for megamammals in the region, Metcalf et al. (2016) did not recover such a time-shift. These authors calibrated the extinction for megamammals in Patagonia at ~12.3 ka, correlated with an abrupt warming phase, environmental changes, and human occupation. In this context, large carnivorans could have gone extinct through direct persecution by humans (cf. Villavicencio et al. 2016) or because humans caused a decline or demise of carnivoran prey (Cione et al. 2009). In the latter case, the competitive displacement and extinction of large carnivorans could have triggered a cascade of secondary extinctions in the ecosystems (cf. Cione et al. 2009). Accordingly, Whitney Smith (2009) proposed the "Second Order Predation Theory," that humans reduced predator populations, leading to a dramatic increase in megaherbivore populations, with a consequent environmental exhaustion and faunal extinction.

Some authors have claimed that the use of domestic dogs could also have contributed to the extinction of these populations under stress, as already hypothesized for Europe and North America (Fiedel 2005; Shipman 2015). However, in South America, the record of domestic dogs dates between 7.5 and 4.5 ka and it is later than the time of the megafaunal extinction (Prates et al. 2010), suggesting that it was not involved in this extinction event.

Unfortunately, there is insufficient evidence from the Late Pleistocene–early Holocene faunas of South America to identify the exact sequence of events (cf. Prevosti and Pereira 2014). Progress has been made with the recent large number of dates published by Metcalf et al. (2016), revealing the simultaneity of the extinction of several megamammals and large predators in southern Patagonia. A conjunction of factors, such as an abrupt change in the climatic and environmental conditions and human presence (along any number of vectors, including hunting pressure, decreasing habitat range, diseases, importation of exotic animals), could have pushed larger mammals to a tipping point leading to their sudden extinction (MacPhee and Marx 1997; Cione et al. 2009; Van Valkenburgh et al. 2015; Metcalf et al. 2016; Fig. 6.9). In this context, the independent role of large carnivores is not clear or discernible from the available data.

After the Late Pleistocene–Holocene extinction, the carnivoran communities of South America survived nearly intact, at least in species composition at a continental scale, except for the extinction of the *Dusicyon* lineage. The last representatives of this genus were the Malvinas fox *D. australis*, which was hunted to extinction during the late nineteenth century, and *D. avus*, which lived on the continent. It disappeared during the last five hundred years, probably due to a combination of environmental change and human impact (Prevosti et al. 2015; Fig. 6.9).

References

Agnolin F (2009) Sistemática y Filogenia de las Aves Fororracoideas (Gruiformes, Cariamae). Fundación de Historia Natural Félix de Azara, Buenos Aires

Alberdi MT, Ortiz Jaureguizar E, Prado JL (1995) Evolución de las comunidades de mamíferos del Cenozoico superior de la Provincia de Buenos Aires, Argentina. Rev Esp Paleontol 10:30–36

Albino AM (1996) The South American fossil Squamata (Reptilia: Lepidosauria). In: Arratia G (ed) Contributions of Southern South America to vertebrate paleontology. Verlag Dr. Friedrich Pfeil, München, pp 185–202

Alvarenga HMF, Höfling E (2003) Systematic revision of the Phorusrhacidae (Aves: Ralliformes). Pap Avulsos Zool 43:55–91

Alvarenga H, Washington J, Rinderknecht A (2009) The youngest record of phorusrhacid birds (Aves, Phorusrhacidae) from the Late Pleistocene of Uruguay. Neues Jahrb Geol Paläontol Abh 256:229–234

Argot C (2003a) Functional adaptations of the postcranial skeleton of two Miocene borhyaenoids (Mammalia, Metatheria) *Borhyaena* and *Prothylacynus*, from South America. Palaeontology 46:1213–1267

Argot C (2003b) Postcranial functional adaptations in the South American Miocene borhyaenoids (Mammalia, Metatheria): *Cladosictis, Pseudonotictis,* and *Sipalocyon.* Alcheringa 27:303–356

Argot C (2004a) Functional-adaptative analysis of the postcranial skeleton of a Laventan borhyaenoid, *Lycopsis longirostris* (Marsupialia, Mammalia). J Vertebr Paleontol 24:689–708

Argot C (2004b) Evolution of South American mammalian predators (Borhyaenoidea): anatomical and palaeobiological implications. Zool J Linn Soc 140:487–521

Argot C (2004c) Functional-adaptive features and paleobiologic implications of the postcranial skeleton of the late Miocene sabretooth borhyaenoid, *Thylacosmilus atrox* (Metatheria). Alcheringa 28:229–266

Babot MJ, Ortiz PE (2008) Primer registro de Borhyaenoidea (Mammalia, Metatheria, Sparassodonta) en la provincia de Tucumán (Formación India Muerta, Grupo Choromoro; Mioceno tardío). Acta Geol Lilloana 21:34–48

Barnosky AD (2001) Distinguishing the effects of the Red Queen and Court Jester on Miocene mammal evolution in the Northern Rocky Mountains. J Vert Pal 21:172–185

Barreda V, Palazzesi L (2007) Patagonian vegetation turnovers during the Paleogene-early Neogene: origin of arid-adapted floras. Bot Rev 73:31–50

Barreda V, Guler V, Palazzesi L (2008) Late Miocene continental and marine palynological assemblages from Patagonia. Dev Quat Sci 11:343–350

Bartlett L, Bartlett DR, Williams GW, Prescott AB, Rhys EG, Eriksson A, Valdes PJ, Singarayer JS, Manica A (2016) Robustness despite uncertainty: regional climate data reveal the dominant role of humans in explaining global extinctions of late Quaternary megafauna. Ecography 39:152–161

Benjamin Y, Hochberg Y (1995) Controlling the false discovery rate: a practical and powerful approach to multiple testing. J R Statist Soc B 57:289–300

Benton MJ (1983) Large-scale replacements in the history of life. Nature 302:16–17

Bocherens H, Cotte M, Bonini R, Scian D, Straccia P, Soibelzon L, Prevosti FJ (2016) Paleobiology of sabretooth cat Smilodon populator in the Pampean Region (Buenos Aires Province, Argentina) around the Last Glacial Maximum: Insights from carbon and nitrogen stable isotopes in bone collagen. Palaeogeogr Palaeoclimatol Palaeoecol 449:463–474

Bond M (1986) Los carnívoros terrestres fósiles de Argentina: resumen de su historia. Actas IV Congr Arg Paleontol Bioest 2:167–171

Bond M, Pascual R (1983) Nuevos y elocuentes restos craneanos de Proborhyaena gigantea Ameghino 1897 (Marsupialia, Borhyanidae, Proborhyaeninae) de la edad Deseadense. Un ejemplo de coevolución. Ameghiniana 20:47–60

Bond M, Cerdeño E, López G (1995) Los ungulados nativos de América del Sur. In: Alberdi MT, Leone G, Tonni EP (eds) Monografías del Museo Nacional de Ciencias Naturales, Madrid, pp 259–275

Borrero LA (2009) The elusive evidence: the archeological record of the South American extinct megafauna. In: Haynes G (ed) American megafaunal extinctions at the end of the Pleistocene, pp 145–168

Brusatte S, Benton M, Ruta M, Lloyd G (2008) Superiority, competition, and opportunism in the evolutionary radiation of dinosaurs. Science 321:1485–1488

Campbell KE, Frailey CD, Romero-Pittman L (2006) The Pan-Amazonian Ucayali peneplain, late Neogene sedimentation in Amazonia, and the birth of the modern Amazon River system. Palaeogeog Palaeoclimatol Palaeoecol 239:166–219

Cardonatto MC, Melchor RN, Mendoza Belmontes FR, Montalvo CI (2016) Large mammal burrows in late Miocene calcareous paleosols from central Argentina. In: Fourth International Congress on Ichnology, Indanha-a-Nova (Portugal), Abstracts

Cassini GH, Cerdeño E, Villafañe AL, Muñoz NA (2012) Paleobiology of Santacrucian native ungulates (Meridiungulata: Astrapotheria, Litopterna and Notoungulata). In: Vizcaíno SF, Kay RF, Bargo MS (eds) Early Miocene Paleobiology in Patagonia. Cambridge University Press, Cambridge, pp 243–286

Cione AL, Tonni EP (2001) Correlation of Pliocene to Holocene southern South American and European vertebrate-bearing units. Boll Soc Paleontol Ital 40:1–7

Cione AL, Tonni EP, Soibelzon LH (2003) The Broken Zig-Zag: late Cenozoic large mammal and turtle extinction in South America. Revista del Museo Argentino de Ciencias Naturales "Bernardino Rivadavia" 5:1–19

Cione AL, Tonni EP, Soibelzon LH (2009) Did humans cause large mammal Late Pleistocene-Holocene extinction in South America in a context of shrinking open areas? In: Haynes G (ed) American megafaunal extinctions at the end of the Pleistocene. Springer Publishers, Vertebrate Paleobiology and Paleontology Series, pp 125–144

Clark PJ, Evans FC (1954) Generalization of a Nearest Neighbormeasure of dispersion for use in K dimensions. Ecology 60:316–317

Cozzuol MA (2006) The Acre vertebrate fauna: age, diversity, and geography. J South Am Earth Sci 21:185–203

Croft DA (2001) Cenozoic environmental change in South America as indicated by mammalian body size distributions (cenograms). Divers Distrib 2:271–287

Croft DA (2006) Do marsupials make good predators? Insights from predator-prey diversity ratios. Evol Ecol Res 8:1193–1214

Dawkins R, Krebs JR (1979) Arms races between and within species. Proc R Soc 205(1161):489–511

Degrange F (2012) Morfología del cráneo y complejo apendicular posterior de aves fororracoideas: implicancias en la dieta y modo de vida. Unpublished PhD thesis, Universidad Nacional de La Plata, La Plata.

Degrange FJ, Noriega JI, Areta JI (2012) Diversity and paleobiology of the Santacrucian birds. In: Vizcaíno SF, Kay RF, Bargo MS (eds) Early Miocene paleobiology in Patagonia. Cambridge University Press, Cambridge, pp 138–155

Degrange FJ, Tambussi CP, Taglioretti ML, Dondas A, Scaglia F (2015) A new Mesembriornithinae (Aves, Phorusrhacidae) provides new insights into the phylogeny and sensory capabilities of terror birds. J Vertebr Paleontol 35(2):e912656. doi:10.1080/02724634.2014.912656

de Muizon C (1994) A new carnivorous marsupial from the Palaeocene of Bolivia and the problem of marsupial monophyly. Nature 370:208–211

de Muizon C (1998) *Mayulestes ferox*, a borhyaenoid (Metatheria, Mammalia) from the early Palaeocene of Bolivia. Phylogenetic and palaeobiologic implications. Geodiversitas 20:19–142

de Muizon C, Cifelli RL, Céspedes Paz R (1997) The origin of the dog-like borhyaenoid marsupials of South America. Nature 389:486–489

de Muizon C, Billet G, Argot C, Ladevèze S, Goussard F (2015) *Alcidedorbignya inopinata*, a basal pantodont (Placentalia, Mammalia) from the early Palaeocene of Bolivia: anatomy, phylogeny and palaeobiology. Geodiversitas 37:397–634

Dickman CR (1986) An experimental study of competition between two species of dasyurid marsupials. Ecol Monog 56:221–241

Donadio E, Buskirk SW (2006) Diet, morphology, and interspecific killing in Carnivora. Am Nat 167:524–536

Dondas A, Isla FI, Carballido J (2009) Paleocaves exhumed from the Miramar Formation (Ensenadan Stage-age). Quat Int 210:44–50

Dozo MT, Bouza P, Monti A, Palazzesi L, Barreda V, Massaferro G, Scasso R, Tambussi C (2010) late Miocene continental biota in northeastern Patagonia (Península Valdés, Chubut, Argentina). Palaeogeogr Palaeoclimatol Palaeoecol 297:100–109

De Elorriaga EE, Visconti G (2002) Crotovinas atribuibles a grandes mamíferos del Cenozoico en el sureste de la Provincia de La Pampa. 9° Reunión Argentina de Sedimentología, Resúmenes 1:63

Engelman RK, Croft DA (2014) A new species of small-bodied sparassodont (Mammalia: Metatheria) from the middle Miocene locality of Quebrada Honda, Bolivia. J Vert Paleontol 34:672–688

Ercoli MD, Prevosti FJ, Álvarez A (2012) Form and function within a phylogenetic framework: locomotory habits of extant predators and some Miocene Sparassodonta (Metatheria). Zool J Linn Soc 165:224–251

Ercoli MD, Prevosti FJ, Forasiepi AM (2014) The structure of the mammalian predator guild in the Santa Cruz Formation (late early Miocene), Patagonia, Argentina. J Mammal Evol 21:369–381

Fariña RA (1996) Trophic relationships among Lujanian mammals. Evol Theo 11:125–134

Fariña RA, Vizcaíno SF, De Iuliis G (2013) Megafauna. Giant Beasts of Pleistocene South America. Indiana University Press, Bloomington, p 416

Fariña RA, Czerwonogora ADA, Giacomo MDI (2014) Splendid oddness: revisiting the curious trophic relationships of South American Pleistocene mammals and their abundance. An Acad Bras Cienc 86:311–331

Faurby S, Svenning, J-C (2016) The asymmetry in the Great American Biotic Interchange in mammals is consistent with differential susceptibility to mammalian predation. Global Ecology and Biogeography 25:1443–1453

Fiedel SJ (2005) Man's best friend—mammoth's worst enemy? A speculative essay on the role of dogs in Paleoindian colonization and megafaunal extinction. World Archaeol 37(1):11–25

Figueirido B, Soibelzon LH (2009) Inferring paleoecology in extinct tremarctine bears (Carnivora, Ursidae) via geometric morphometrics. Lethaia 43:209–222

Finarelli JA, Liow LH (2016) Diversification histories for North American and Eurasian carnivorans. Biol J Linn 118(1):26–38

Forasiepi AM (2009) Osteology of *Arctodictis sinclairi* (Mammalia, Metatheria, Sparassodonta) and phylogeny of Cenozoic metatherian carnivores from South America. Monogr Mus Argent Cienc Nat 6:1–174

Forasiepi AM, Carlini AA (2010) New thylacosmilid (Mammalia, Metatheria, Sparassodonta) from the Miocene of Patagonia, Argentina. Zootaxa 2552:55–68

Forasiepi AM, Goin FJ, Tauber AA (2004) Las especies de *Arctodictis* Mercerat, 1891 (Metatheria, Borhyaenidae), grandes metaterios carnívoros del Mioceno de América del Sur. Rev Esp Paleontol 19:1–22

Forasiepi AM, Martinelli AG, Goin FJ (2007) Revisión taxonómica de *Parahyaenodon argentinus* Ameghino y sus implicancias en el conocimiento de los grandes mamíferos carnívoros del Mio-Plioceno de América del Sur. Ameghiniana 44:143–159

Forasiepi AM, Sánchez-Villagra MR (2014) Heterochrony, dental ontogenetic diversity and the circumvention of constraints in marsupial mammals and extinct relatives. Paleobiology 40:222–237

Forasiepi AM, Babot MJ, Zimicz N (2015) *Australohyaena antiqua* (Mammalia, Metatheria, Sparassodonta), a large predator from the late Oligocene of Patagonia, Argentina. J Syst Palaeontol 13:503–525

Forasiepi AM, Macphee RDE, Hernández Del Pino S, Schmidt GI, Amson E, Grohe C (2016) Exceptional skull of Huayqueriana (Mammalia, Litopterna, Macraucheniidae) from the late Miocene of Argentina: anatomy, systematics, and paleobiological implications. Bulletin of the American Museum of Natural History 404:1–76

Friscia A VanValkenburgh B (2010) Ecomorphology of North American Eocene carnivores: evidence for competition between carnivorans and creodonts. In: Goswami A, Friscia A (eds) Carnivoran evolution: new views on phylogeny, form, and function. Cambridge University Press, Cambridge and New York, pp 311–341

Gasparini Z (1996) Biogeographic evolution of the South American crocodilians. Münchner Geowiss Abh 30:159–184

Genise JF, Melchor RN, Sánchez MV, González MG (2013) *Attaichnus kuenzelii* revisited: a Miocene record of fungus-growing ants from Argentina. Palaeogeogr Palaeoclimatol Palaeoecol 386:349–363

Goin FJ (1989) Late Cenozoic South American marsupial and placental carnivores: changes in predator-prey evolution. V Internatl Ther Congr, Abstracts: 271–272

Goin FJ (1995) Los marsupiales. In: Alberdi MT, Leone G, Tonni EP (eds) Evolución biológica y climática de la Región Pampeana durante los últimos cinco millones de años. Un ensayo de correlación con el Mediterráneo Occidental, Monografías del Museo Nacional de Ciencias Naturales, Madrid, pp 165–179

Goin FJ, Pascual R (1987) News on the biology and taxonomy of the marsupials Thylacosmilidae (late Tertiary of Argentina). Acad Nac Cienc Exactas Fis Nat B Aires 39:219–246

Goin FJ, Pardiñas UFJ (1996) Revisión de las especies del género *Hyperdidelphys* Ameghino, 1904 (Mammalia, Marsupialia, Didelphidae), su significado filogenético, estratigráfico y adaptativo en el Neógeno del Cono Sur Sudamericano. Estudios Geol 52:327–359

Goin FJ, Forasiepi AM, Candela AM, Ortiz-Jaureguizar E, Pascual R, Archer M, Godthelp H, Muirhead J, Augee M, Hand S, Wroe S (2002) Earliest Paleocene Marsupials from Patagonia. In: First International Palaeontological Congress. Sydney, Abstracts

Goin FJ, Abello MA, Chornogubsky L (2010) Middle tertiary marsupials from central Patagonia (early Oligocene of Gran Barranca): understanding South America's Grande Coupure. In: Madden RH, Carlini AA, Vucetich MG, Kay RF (eds) The paleontology of Gran Barranca: evolution and environmental change through the middle Cenozoic of Patagonia. Cambridge University Press, New York, pp 71–107

Goin FJ, Gelfo NF, Chornogubsky L, Woodburne MO, Martin T (2012) Origins, radiations, and distribution of South American Mammals: from greenhouse to icehouse worlds. In: Patterson BD, Costa LP (eds) Bones, clones, and biomes: an 80-million year history of Recent Neotropical mammals. University of Chicago Press, Chicago, pp 20–50

Goin FJ, Woodburne MO, Zimicz AN, Martin GM, Chornogubsky L (2016) A brief history of South American metatherians, evolutionary contexts and intercontinental dispersals. Springer, Dordrecht

Goswami A, Milne N, Wroe S (2011) Biting through constraints: cranial morphology, disparity and convergence across living and fossil carnivorous mammals. Proc R Soc B 278:1831–1839

Gould S, Calloway C (1980) Clams and brachiopods; ships that pass in the night. Paleobiology 6:383–396

Gould GC, McFadden BJ (2004) Gigantism, dwarfism, and Cope's Rule: "Nothing in evolution makes sense without a phylogeny". Bull Am Mus Natl Hist 285:219–237

Green B (1997) Field energetics and water flux in marsupials. In: Saunders NR, Hinds LA (eds) Marsupial biology: recent research, new perspectives. University of New South Wales Press, Sidney, pp 143–162

Hammer O (2016) Past 3.0. http://folk.uio.no/ohammer/past

Hammer O, Harper DAT, Ryan PD (2001) PAST: Paleontological statistics software package for education and data analysis. Pal Elect 4(1):9 pp

Jones W, Rinderknecht A, Montenegro F, Alvarenga H, Ubilla M (2016) New evidences of terror birds (Cariamae, Phorusrhacidae) from the Late Pleistocene of Uruguay. In: 9th international meeting of the society of Avian paleontology and evolution, abstracts, vol 1, p 16

Kraglievich JL (1952) El perfil geológico de Chapadmalal y Miramar. Resumen Preliminar. Rev Mus Mun Cs Nat y Trad Mar del Plata 1:8–37

Kraglievich JL, Olazábal AG (1959) Los prociónidos extinguidos del género Chapalmalania Ameghino. Rev Mus Argent Cienc Nat, Zool 6:1–59

Krause DW (1986) Competitive exclusion and taxonomic displacement in the fossil record: the case of rodents and multituberculates in North America. In: Flanagan KM, Lillegraven JA (eds) Vertebrates, phylogeny and philosophy. University of Wyoming Contributions to Geology, Special Paper 3:119–130

Krockenberger A (2006) Lactation. In: Armati P, Dickman C, Hume I (eds) Marsupials. Cambridge University Press, Cambridge, pp 108–136

Langer M, Ezcurra M, Bittencourt J, Novas F (2009) The origin and early evolution of dinosaurs. Biol Rev 85:55–110

Latrubesse EM, da Silva SAF, Cozzuol M, Absy ML (2007) Late Miocene continental sedimentation in southwestern Amazonia and its regional significance: biotic and geological evidence. J South Am Earth Sci 23:61–80

Lima-Ribeiro MS, Diniz-Filho JAF (2013) American megafaunal extinctions and human arrival: improved evaluation using a meta-analytical approach. Quat Int 299:38–52

Liow LH, Finarelli JA (2014) A dynamic global equilibrium in carnivoran diversification over 20 million years. Proc R Soc B 281:20132312

López-Aguirre C, Archer M, Hand SJ, Laffan SW (2017). Extinction of South American sparassodontans (Metatheria): environmental fluctuations or complex ecological processes? Palaeontology 60:91–115

MacFadden BJ, Cerling TE, Prado J (1996) Cenozoic terrestrial ecosystem evolution in Argentina: evidence from carbon isotopes of fossil mammal teeth. Palaios 11:319–327

MacFadden BJ, Labs-Hochstein J, Hulbert RC (2007) Revised age of the late Neogene terror bird (Titanis) in North America during the Great American Interchange. Geology 35(2):123–126

MacPhee RDE, Marx PA (1997) The 40,000-year plague: humans, hyperdisease, and first-contact extinctions. Smithsonian Institution Press, Washington, DC, pp 169–217

Marengo H (2015) Neogene micropaleontology and stratigraphy of Argentina. The Chaco-Paranense basin and the Península de Valdés. Springer, Dordrecht

Marshall LG (1976) Evolution of the Thylacosmilidae, extinct sabertooth marsupials of South America. PaleoBios 23:1–30

Marshall LG (1977) Evolution of the carnivorous adaptive zone in South America. In: Hecht MK, Goody PC, Hecht BM (eds) Major patters in vertebrate evolution. Plenum Press, New York, pp 709–722

Marshall LG (1978) Evolution of the Borhyaenidae, extinct South American predaceous marsupials. Univ Calif Publ Geol Sci 117:1–89

Marshall LG (1979) Review of the Prothylacyninae, an extinct subfamily of South American "dog–like" marsupials. Fieldiana. Geol [ns] 3:1–50

Marshall LG (1981) Review of the Hathlyacyninae, an extinct subfamily of South American "dog–like" marsupials. Fieldiana. Geol [ns] 7:1–120

Marshall LG, Cifelli RL (1990) Analysis of changing diversity patterns in Cenozoic Land Mammal Age faunas, South America. Palaeovertebrata 19:169–210

Martin PS, Wright HEJ (eds) (1967) Pleistocene extinctions: the search for a cause. Yale University Press, New Haven

McNab BK (1986) Food habits, energetics and the reproduction of marsupials. J Zool 208:595–614

McNab BK (2005) Uniformity in the basal metabolic rate of marsupials: its causes and consequences. Rev Chil Hist Nat 78:183–219

McNab BK (2008) An analysis of the factors that influence the level and scaling of mammalian BMR. Comp Biochem Physiol A 151:5–28

Metcalf JL, Turney C, Barnett R, Martin F, Bray SC, Vilstrup JT, Orlando L, Salas-Gismondi R, Loponte D, Medina M, De Nigris M, Civalero T, Fernández PM, Gasco A, Duran V, Seymour KL, Otaola C, Gil A, Paunero R, Prevosti FJ, Bradshaw CJA, Wheeler JC, Borrero L, Austin JJ, Cooper A (2016) Synergistic roles of climate warming and human occupation in Patagonian megafaunal extinctions during the Last Deglaciation. Sci Adv 2:e1501682

Mitchell KJ, Bray SC, Bover P, Soibelzon L, Schubert BW, Prevosti FJ, Prieto A, Martin F, Austin JJ, Cooper A (2016) Ancient mitochondrial DNA reveals convergent evolution of giant short-faced bears (Tremarctinae) in North and South America. Biol Let 12:20160062

Oliveira TG, Pereira JA (2014) Intraguild predation and interspecific killing as structuring forces of Carnivoran Communities in South America. J Mam Evol 21(4):427–436

Ortiz Jaureguizar E (1989) Analysis of the compositional changes of the South American mammal fauna during the Miocene-Pliocene (Panaraucanian Faunistic Cycle). V Internatl Ther Congr Abstracts 277–278

Ortiz Jaureguizar E (2001) Cambios en la diversidad de los mamíferos sudamericanos durante el lapso Mioceno Superior-Holoceno: el caso pampeano. In: Meléndez G, Herrera Z, Delvene G, Azanza B (eds) Los fósiles y la paleogeografía. Publicaciones del SEPAZ, Univ Zaragoza 5:397–403

Ortiz Jaureguizar E, Prado JL, Alberdi MT (1995) Análisis de las comunidades de mamíferos continentales del Plio-Pleistoceno de la región pampeana y su comparación con la del área de mediterráneo occidental. In: Alberdi MT, Leone G, Tonni EP (eds) Evolución biológica y climática de la Región Pampeana durante los últimos cinco millones de años. Un ensayo de correlación con el Mediterráneo Occidental. Monografías del Museo Nacional de Ciencias Naturales, Madrid, pp 385–406

Palomares F, Caro TM (1999) Interspecific killing among mammalian carnivores. Am Nat 153:493–508

Paolillo A, Linares OJ (2007) Nuevos cocodrilos Sebecosuchia del Cenozoico Suramericano (Mesosuchia: Crocodylia). Paleobiología Neotropical 3:1–25

Pascual R, Odreman Rivas OE (1971) Evolución de las comunidades de vertebrados del Terciario Argentino. Los aspectos paleozoogeográficos y paleoclimáticos relacionados. Ameghiniana 8:372–412

Pascual R, Bond M (1986) Evolución de los marsupiales cenozoicos de Argentina. Actas IV Congr Arg Paleontol y Bioest 2:143–150

Pascual R, Ortiz Jaureguizar E (1990) Evolving climates and mammal faunas in Cenozoic South American. J Human Evol 19:23–60

Pascual R, Vucetich MG, Scillato-Yané GJ, Bond M (1985) Main pathways of mammalian diversification in South America. In: Stehli FG, Webb SD (eds) The Great American Biotic Interchange. Plenum Press, New York, pp 219–247

Patterson B, Pascual R (1972) The fossil mammal fauna of South America. In: Keast A, Erk FC, Glass B (eds) Evolution, mammals and Southern continents. State University of New York Press, New York, pp 274–309

Pires MM, Silvestro D, Quental TB (2015) Continental faunal exchange and the asymmetrical radiation of carnivores. Proc R Soc B 282:20151952. doi:10.1098/rspb.2015.1952

Politis GG, Gutiérrez MA, Rafuse DJ, Blasi A (2016) The arrival of Homo sapiens into the Southern Cone at 14,000 years ago. PLoS ONE 11(9):e0162870. doi:10.1371/journal.pone.0162870

Prado JL, Martínez-Maza C, Alberdi MT (2015) Megafauna extinction in South America: a new chronology for the Argentine Pampas. Palaeogeogr Palaeoclimatol Palaeoecol 425:41–49

Prates L, Prevosti FJ, Berón M (2010) First records of prehispanic dogs in Southern South America (Pampa, Patagonia, Argentina). Curr Anthropol 51:273–280

Prevosti FJ (2006) Grandes cánidos (Carnivora, Canidae) del Cuaternario de la Republica Argentina: sistemática, filogenia, bioestratigrafíay paleoecología. Unpublished PhD thesis, Universidad Nacional de La Plata, La Plata, 506 pp

Prevosti FJ, Vizcaíno S (2006) The carnivore guild of the late Pleistocene of Argentina: Paleoecology and carnivore richness. Acta Palaeontol Pol 51:407–422

Prevosti FJ, Soibelzon LH (2012) The evolution of South American carnivore fauna: a paleontological perspective. In: Patterson B, Costa LP (eds) Bones, clones and biomes: the history and geography of recent Neotropical mammals. University Chicago Press, Chicago, pp 102–122

Prevosti FJ, Martin FM (2013) Paleoecology of the mammalian predator guild of the southern patagonia during the latest Pleistocene: ecomorphology, stable isotopes, and taphonomy. Quatern Int 305:74–84

Prevosti FJ, Pereira JA (2014) Community structure of South American Carnivores in the past and present. J Mam Evol 21(4):363–368

Prevosti FJ, Turazzini GF, Chemisquy MA (2010) Morfología craneana en tigres dientes de sable: alometría, función y filogenia. Ameghiniana 47:239–256

Prevosti FJ, Forasiepi AM, Ercoli MD, Turazzini GF (2012) Paleoecology of the mammalian carnivores (Metatheria, Sparassodonta) of the Santa Cruz Formation (late early Miocene). In: Vizcaíno SF, Kay RF, Bargo MS (eds) early Miocene paleobiology in Patagonia. Cambridge University Press, Cambridge, pp 173–193

Prevosti FJ, Forasiepi A, Zimicz N (2013) The evolution of the Cenozoic terrestrial mammalian predator guild in South America: competition or replacement? J Mammal Evol 20:3–21

Prevosti FJ, Ramírez MA, Martin F, Udrizar Sauthier DE, Carrera M (2015) Extinctions in near time: new radiocarbon dates point to a very recent disappearance of the South American fox Dusicyon avus (Carnivora: Canidae) Biol J Linn Soc 116:704–720

Raia P, Piras P, Kotsakis T (2005) Turnover pulse or Red Queen? Evidence from the large mammal communities during the Plio-Pleistocene of Italy. Palaeogeogr Palaeoclimatol Palaeoecol 221:293–312

Reig OA (1981) Teoría del origen y desarrollo de la fauna de mamíferos de América del Sur. Monographiae Naturae 1:1–162

Reguero MA, Candela A, Cassini G (2010) Hypsodonty and Body Size in rodent-like notoungulates. The Paleontology of Gran Barranca: Evolution and Environmental Change through the Middle Cenozoic of Patagonia. New York, Cambridge University Press, 1–448

Riff D, Romano PSR, Ribeiro Oliveira G, Aguilera AO (2010) Neogene crocodile and turtle fauna. In: Hoorn C, Wesselingh FP (eds) Northern South America Amazonia, landscape and species evolution: a look into the past. Blackwell Publishing, Oxford, pp 259–280

Rosenzweig ML, McCord RD (1991) Incumbent replacement: evidence for long-term evolutionary progress. Paleobiology 17:202–213

Rutter N, Coronato A, Helmens K et al (2012) Glaciations in North and South America from the Miocene to the Last Glacial Maximum. Comparisons, linkages and uncertainties. Springer, Dordrecht

Savage RJG (1977) Evolution in carnivorous mammals. Palaeontology 20:237–271

Schultz PH, Zarate M, Hames W, Camilión C, King J (1998) A 3.3 Ma impact in Argentina and possible consequences. Science 282:2061–2063

Segura V, Cassini GH, Prevosti FJ (2016) Three-dimensional cranial ontogeny in pantherines (*Panthera leo*, *P. onca*, *P. pardus*, *P. tigris*; Carnivora: Felidae). Biol J Linnean. doi:10.1111/bij.12888

Sepkoski JJ Jr (1996) Patterns of Phanerozoic extinctions: a perspective from global databases. In: Walliser H (ed) Global events and event stratigraphy. Springer, Berlin, pp 35–52

Sepkoski JJ Jr (2001) Competition in evolution. In: Briggs DEG, Crowther PR (eds) Paleobiology II. Blackwell Sciences, Oxford, pp 171–175

Sepkoski JJ Jr, McKinney FK, Lidgard S (2000) Competitive displacement among post-Paleozoic cyclostome and cheilostome bryozoans. Paleobiology 26:7–18

Shipman P (2015) The Invaders: how humans and their Dogs Drove Neanderthals to Extinction. Harvard university Press, Cambridge (Massachusetts), pp 1–266

Silvestro D, Antonelli A, Salamin N, Quental TB (2015) The role of clade competition in the diversification of North American canids. PNAS 112(28):8684–8689

Simpson GG (1950) History of the fauna of Latin America. Am Scientist 38:361–389

Simpson GG (1969) South American mammals. In: Fitkau EJ, Illies J, Klinge H, Schwabe GH, Sioli H (eds) Biogeography and ecology in South America. Dr. W Junk Publishers, The Hague, pp 876–909

Simpson GG (1971) The evolution of marsupials in South America. An Acad Bras Cs 43:103–118

Simpson GG (1980) Splendid isolation. The curious history of South American mammals. Yale University Press, New Haven

Soibelzon LH (2011) First description of milk teeth of fossil South American procyonid from the lower Chapadmalalan (late Miocene-early Pliocene) of "Farola Monte Hermoso", Argentina: paleoecological considerations. Palaontologische Z 85:83–89

Soibelzon LH, Schubert BW (2011) The largest known bear, *Arctotherium angustidens*, from the early Pleistocene pampean region of Argentina: with a discussion of size and diet trends in bears. J Paleontol 85:69–75

Soibelzon E, Gasparini GM, Zurita AE, Soibelzon LH (2008) Análisis faunístico de vertebrados de las toscas del Río de La Plata (Buenos Aires, Argentina): un yacimiento paleontológico en desaparición. Rev Mus Arg Cs Nat 10(2):291–308

Soibelzon LH, Pomi LH, Tonni EP, Rodriguez S, Dondas A (2009) First report of a South American short-faced bears' den (*Arctotherium angustidens*): palaeobiological and palaeoecological implications. Alcheringa 33(3):211–222

Soibelzon LH, Zamorano M, Scillato-Yané GJ, Soibelzon E, Tonni EP (2012) Un glyptodontidae de gran tamaño en el holoceno temprano de la región pampeana, argentina. Rev Bras Paleontol 15(1):105–112

Steele J, Politis G (2009) AMS ^{14}C dating of early human occupation of southern South America. J Arch Sci 36(2):419–429

Suarez C, Forasiepi AM, Goin FJ, Jaramillo C (2015) Insights into the Neotropics prior to the Great American Biotic Interchange: new evidence of mammalian predators from the Miocene of Northern Colombia. J Vert Paleontol. doi:10.1080/02724634.2015.1029581

Surovell TA, Pelton SR, Anderson-Sprecher R, Myers AD (2015) Test of Martin's overkill hypothesis using radiocarbon dates on extinct megafauna. PNAS 113(4):886–891

Tambussi C, Ubilla M, Perea D (1999) The youngest large carnassials bird (Phorusrhacidae, Phorusrhacinae) from South America (Pliocene–early Pleistocene of Uruguay). J Vertebr Paleontol 19:404–406

Tonni EP (1977) El rol ecológico de algunas aves fororracoideas. Ameghiniana 14:316

Tonni EP, Alberdi MT, Prado JL, Bargo MS, Cione AL (1992) Changes of mammal assemblages in the Pampean Region (Argentina) and their relation with the Plio-Pleistocene boundary. Palaeogeogr Palaeoclimatol Palaeoecol 95:179–194

Van Valen L (1973) A new evolutionary law. Evol Theor 1:1–30

Van Valkenburgh B (1999) Major patterns in the history of carnivorous mammals. Ann Rev Earth Plan Sci 27:463–493

Van Valkenburgh B (2007) Déjà vu: the evolution of feeding morphologies in the Carnivora. Integr Comp Biol 47:147–163

Van Valkenburgh B, Wang X, Damuth J (2004) Cope's rule, hypercarnivory, and extinction in North American canids. Science 306:101–104

Van Valkenburgh B, Hayward MW, Ripple WJ, Meloro C, Roth VL (2015) The impact of large terrestrial carnivores on Pleistocene ecosystems. PNAS 113(4):862–867

Villafañe AL, Ortiz-Jaureguizar E, Bond M (2006) Cambios en la riqueza taxonómica y en las tasas de primera y última aparición de los Proterotheriidae (Mammalia, Litopterna) durante el Cenozoico. Estudios Geológicos 62:155–166

Villavicencio NA, Lindsey EL, Martin FM, Borrero LA, Moreno PI, Marshall CR, Barnosky AD (2016) Combination of humans, climate, and vegetation change triggered late Quaternary megafauna extinction in the Última Esperanza region, southern Patagonia, Chile. Ecography 39:125–140

Viranta S (1996) European Miocene Amphicyonidae—taxonomy, systematics and ecology. Acta Zool Fen 204:1–61

Vizcaíno SF, Bargo M, Dondas A (2001) Pleistocene burrows in the Mar del Plata area (Argentina) and their probable builders. Acta Pal Pol 46(2):289–301

Vizcaíno SF, Fariña RA, Zárate MA, Bargo MS, Schultz P (2004) Palaeoecological implications of the mid-Pliocene faunal turnover in the Pampean Region (Argentina). Palaeogeogr Palaeoclimatol Palaeoecol 213:101–113

Vizcaíno SF, Cassini GH, Toledo N, Bargo MS (2012) On the evolution of large size in Mammalian Herbivores of Cenozoic faunas of Southern South America. In: Patterson B, Costa LP (eds) Bones, clones and biomes: the history and geography of Recent Neotropical mammals. University Chicago Press, Chicago, pp 76–101

Wang X, Tedford R, Antón M (2008) Dogs: their fossil relatives and evolutionary history. Columbia University Press, Chichester

Werdelin L (1987) Jaw geometry and molar morphology in marsupial carnivores: analysis of a constraint and its macroevolutionary consequences. Paleobiology 13:342–350

Whitney Smith E (2009) The second-order predation hypothesis of pleistocene extinctions: a system dynamics model. VDM Verlag Dr. Müller, Saarbrücken, pp 1–156

Wroe S, Argot C, Dickman C (2004) On the rarity of big fierce carnivores and primacy of isolation and area: tracking large mammalian carnivore diversity on two isolated continents. Proc Royal Soc Lond 271:1203–1211

Wroe S, Chamoli U, Parr WCH (2013) Comparative Biomechanical Modeling of Metatherian and Placental Saber-Tooths: A Different Kind of Bite for an Extreme Pouched Predator. PLoS One 8:8–16

Zachos J, Pagani M, Sloan L, Thomas E, Billups K (2001) Trends, rhythms, and aberrations in global climate 65 Ma to present. Science 292:686–693

Zimicz AN (2012) Ecomorfología de los marsupiales paleógenos de América del Sur. Unpublished Ph.D. thesis, Universidad Nacional La Plata

Zimicz AN (2014) Avoiding competition: the ecological history of metatherian carnivores. J Mamm Evol 21:383–393

Zurita AE, Soibelzon L, Soibelzon E, Gasparini GM, Cenizo M, Arzani H (2010) Accessory protection structures in Glyptodon Owen (Xenarthra, Cingulata, Glyptodontidae). Ann Pal 96(1):1–11